MSXR209 Mathe

Introduction to mathematical modelling

Study guide

This course text introduces the study of mathematical modelling. The *Course Guide* indicates what background knowledge is required for successful study of the course, including this text.

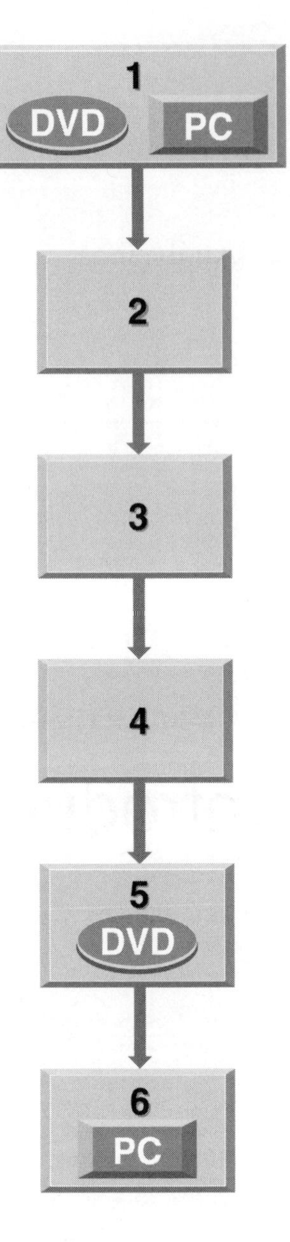

You should study the material in the order in which it is presented. Section 1, which relates to a major case study, includes both a video sequence and a multimedia package, as well as reading matter. If possible, you should complete all aspects of the section in a single session of four hours. Otherwise, try to study the video, multimedia and text in fairly rapid succession, in order to gain as much as possible from the case study.

Section 2 asks you to analyse a previously formulated model. This should take about three hours to study, and again is best done in one session.

In Section 3 you are asked to tackle a small modelling problem, which should take about three hours. This develops the various stages in the modelling process, and is a good guide for the types of activity that you will be asked to undertake for the course assessment.

Section 4 introduces another major case study, and should take about three hours to study.

In Section 5, which should take about three hours to study, the case study from Section 1 is evaluated and then revised. There is a further video sequence associated with this section.

Section 6 should take about four hours to study. It shows how to use the computer algebra package associated with the course to solve, interpret and evaluate the models. This is followed by a final multimedia session.

This publication forms part of an Open University course. Details of this and other Open University courses can be obtained from the Student Registration and Enquiry Service, The Open University, PO Box 625, Milton Keynes, MK7 6YG, United Kingdom: tel. +44 (0)1908 653231, e-mail general-enquiries@open.ac.uk

Alternatively, you may visit the Open University website at http://www.open.ac.uk where you can learn more about the wide range of courses and packs offered at all levels by The Open University.

To purchase a selection of Open University course materials, visit the webshop at www.ouw.co.uk, or contact Open University Worldwide, Michael Young Building, Walton Hall, Milton Keynes, MK7 6AA, United Kingdom, for a brochure: tel. +44 (0)1908 858785, fax +44 (0)1908 858787, e-mail ouwenq@open.ac.uk

The Open University, Walton Hall, Milton Keynes, MK7 6AA.

First published 2005.

Edited, designed and typeset by The Open University, using the Open University TEX System.

Printed and bound in the United Kingdom by Thanet Press Ltd, Margate.

ISBN 0 7492 0298 X

1.1

Contents

Introduction

Applied mathematics is concerned not only with the development of *mathematical methods* but also with the application of these methods in *mathematical models*, which are idealized representations of aspects of the real world. This text aims to develop your appreciation of the role of mathematics in understanding and predicting the behaviour of the real world (as opposed to the world of mathematical theories). This process is called *mathematical modelling*.

For example, some mathematical methods are dealt with in MST209 *Units 1–4*, whereas an established mathematical model is the subject of MST209 *Units 5–8*.

There was a time when the application of mathematics was restricted more or less to physics and engineering. Those days are long gone, and now the use of mathematics is widespread. This *Introduction to mathematical modelling* introduces skills that will enable you to develop your own mathematical models for simple real-world situations. The *mathematical modelling process* starts with a problem in the 'real world'. This problem is translated into a mathematical model, whose solution may provide solutions to the original real-world problem. The mathematical model may also help to predict what will happen in the real world if changes are made. The key stages in the mathematical modelling process are as follows.

This process was described in MST121 *Chapter A1*, and also in MST209 *Unit 16*.

1 **Specify the purpose of the model:**
define the problem;
decide which aspects of the problem to investigate.

2 **Create the model:**
state assumptions;
choose variables and parameters;
formulate mathematical relationships.

3 **Do the mathematics:**
solve equations;
draw graphs;
derive results.

4 **Interpret the results:**
collect relevant data;
describe the mathematical solution in words;
decide what results to compare with reality.

5 **Evaluate the model:**
test the model by comparing its predictions with reality;
criticize the model.

The diagram in Figure 0.1 may help you to remember the five key stages in the mathematical modelling process.

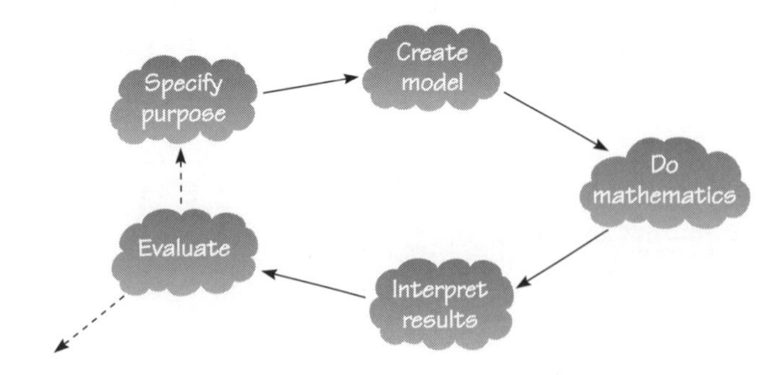

Figure 0.1 The mathematical modelling process (or cycle)

The mathematical modelling process is sometimes referred to as the *mathematical modelling cycle*.

In developing a mathematical model, you may need to go around the loop in Figure 0.1 several times, improving your model each time. When you start to create a mathematical model for a real problem, begin with a simple model, in order to obtain a feel for the problem. Often a simple model that gives a reasonable approximation to the real world is more useful than a complicated model. A more complex model may give a better fit to the available data, but the key processes that are being modelled may be more difficult to identify, and the mathematical problem may be harder to solve. If, in evaluating a model, you find that it is not satisfactory for its purpose, then identify why it is deficient and try to include additional features that address those deficiencies in your next pass through the modelling loop.

As you become more experienced as a mathematical modeller, you will probably find that the above schema for mathematical modelling is insufficiently flexible for the mathematical model that you are developing. The purpose of specifying the modelling process clearly is to try to bring some structure to the subject, but the schema should not be seen as a straitjacket. However, it is recommended that you use this schema when tackling the modelling problems that you will encounter in this course.

Mathematical modelling is unlike other branches of mathematics. The mathematical models that you develop depend critically on the original problem and on the simplifying assumptions that you make about the system being modelled. Consequently, there are many ways of tackling any modelling problem and none can be expected to yield an exact solution. In attempting the modelling exercises in this text, your mathematical model may differ from the one suggested in the solutions, but provided that your model is justified by reasoned arguments and assumptions, it may be just as valid as the one developed here.

1 Pollution in the Great Lakes

This section explores a real-world system where mathematical modelling has been used to understand what is happening and to predict what will happen if changes are made. The system concerned is extremely complex but, by keeping things as simple as possible, sufficient information will be extracted to allow a mathematical model of the system to be obtained. Refinements to this simple model will be made in Section 5.

The Great Lakes of North America (Figure 1.1) provide drinking water for tens of millions of people who live in the surrounding area. They also provide a source of food, transport and recreation. In the first half of the twentieth century they were used for dumping sewage and other pollutants. Sources of pollution include industrial waste, agricultural chemicals, acid rain, and oil and chemical spills. The case study concerns the construction of a mathematical model of how the pollution level in a lake varies over time.

The information for this case study comes from two main sources:

Rainey, R.H. (1967) 'Natural displacement of pollution from the Great Lakes', *Science*, **155**, 1242–3;

Thomann, R.V. and Mueller, J.A. (1987) *Principles of Surface Water Quality Modeling and Control*, Harper and Row.

Figure 1.1 The Great Lakes of North America

Now view the first video sequence for this text, 'Pollution in the Great Lakes'.

DVD

Then work through the first multimedia package, 'Modelling in the Great Lakes'.

Summary of the modelling stages

The multimedia package discussed in detail how the stages of the mathematical modelling process can be applied to the problem of predicting future pollution levels in the Great Lakes. A summary of these details is given below for reference. This summary also includes some information from the video and some additional material not included in the package or on the video.

1 Specify the purpose of the model

Define the problem

The problem is to predict how long it will take for the level of pollution in a lake to reduce to a target level if all sources of pollution are eliminated. It is intended to use this mathematical model to investigate pollution levels in any one of the Great Lakes, although the model could be used for any polluted lake.

Decide which aspects of the problem to investigate

The model investigates how the pollution level varies with time as clean water flows into the lake and polluted water flows out.

2 Create the model

State assumptions

The model assumes that:

(a) all sources of pollution have ceased;

(b) the pollutant does not biodegrade in the lake or decay through any other biological, chemical or physical process;

(c) the pollutant is evenly dispersed within the lake at all times;

(d) water flows into and out of the lake at the same constant rate (so that all seasonal effects can be ignored);

(e) all other water gains and losses (e.g. rainfall, evaporation, extraction and seepage) can be ignored;

(f) the volume of water in the lake is constant;

(g) if the mathematical model is to be used for the downstream lakes (Huron, Erie and Ontario), then negligible pollution is flowing into them from the upstream lakes.

The statement that the volume is constant is a consequence of the previous two assumptions, but it is included as an assumption in its own right because of its importance in the modelling process.

Choose variables and parameters

The variables in the model are:

t — the time, in seconds, since all sources of pollution ceased;

$m(t)$ — the mass, in kilograms, of pollutant in the lake at time t;

$c(t)$ — the concentration, in kilograms per cubic metre, of pollutant in the lake at time t.

The parameters in the model are:

c_{target} — the target concentration level, in kilograms per cubic metre, of pollutant in the lake;

T — the time taken, in seconds, to reduce the concentration of pollutant to the target level c_{target} (that is, $c(T) = c_{\text{target}}$);

V — the volume, in cubic metres, of water in the lake;

r — the water flow rate, in cubic metres per second, into and out of the lake;

$k = r/V$ — the proportionate flow rate, in seconds^{-1}.

The essential difference between a variable and a parameter is that a variable is a quantity whose values change during the situation the model describes, while a parameter is a constant of the model (for a given situation).

Formulate mathematical relationships

The relationship between concentration and mass is given by

$$c(t) = \frac{m(t)}{V}.$$

The **input–output principle**,

$$\boxed{\text{accumulation}} = \boxed{\text{input}} - \boxed{\text{output}},$$

is applied to the mass of pollutant during the time interval $[t, t + \delta t]$. This principle is often used in mathematical modelling. You will see another example of its use in Section 4.

The input–output principle is also applied in MST209 *Unit 16*.

By Assumptions (a) and (g), the mass of pollutant entering the lake is zero, so the *input* is zero. By Assumption (b), the pollutant leaves the lake only through the outflow of water. In the time interval $[t, t + \delta t]$, the volume of water leaving the lake is $r\,\delta t$, where r is constant because of Assumptions (d), (e), and (f). Multiplying this by the concentration of pollutant, which is uniform by Assumption (c), gives the mass of pollutant that leaves the lake in that time interval as $rc(t)\delta t$ or $(r/V)m(t)\delta t$. This is the *output*.

The *accumulation* of the mass of pollutant within the lake, over the time interval $[t, t + \delta t]$, is the difference between $m(t + \delta t)$ and $m(t)$, that is, $m(t + \delta t) - m(t)$. Applying the input–output principle therefore gives

$$m(t + \delta t) - m(t) \simeq 0 - \frac{r}{V} m(t)\delta t.$$

It is sometimes useful to replace a group of parameters by a single parameter, particularly when the same group of parameters is often repeated; this is called **reparametrization**. Replacing r/V by k, dividing by δt and then letting $\delta t \to 0$, leads to the differential equation

$$\frac{dm}{dt} = -km(t), \quad \text{where } k = \frac{r}{V}. \tag{1.1}$$

3 Do the mathematics

Solve equations

The solution to the differential equation for m is

$$m(t) = m(0)e^{-kt}, \quad \text{where } m(0) \text{ is the initial mass of pollutant.}$$

Since $c(t) = m(t)/V$, the pollutant concentration c is given by

$$c(t) = c(0)e^{-kt}, \quad \text{where } c(0) \text{ is the initial concentration of pollutant.}$$

The time, T, taken for the pollutant concentration to reduce to the target level, $c_{\text{target}} = c(T) = c(0)e^{-kT}$, is given by

$$T = -\frac{1}{k} \ln\left(\frac{c_{\text{target}}}{c(0)}\right).$$

Draw graphs

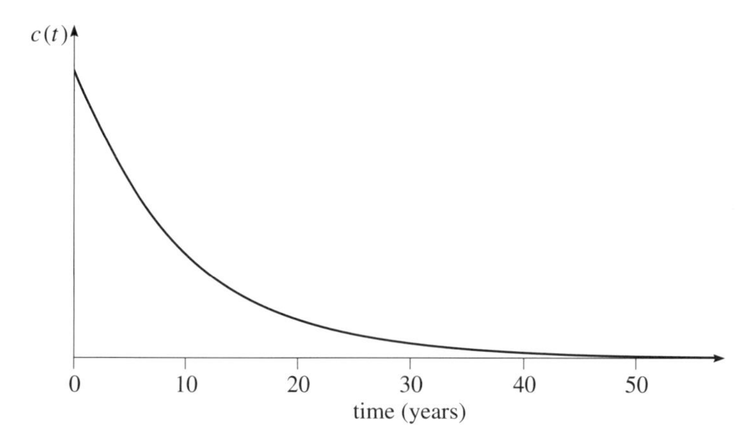

Figure 1.2 Typical graph of concentration against time

Derive results

Suppose that a political decision has been taken to reduce the level of pollution to a tenth of its initial level, so that $c_{\text{target}}/c(0) = \frac{1}{10}$. The corresponding length of time T required is

$$T = -\frac{1}{k}\ln\left(\frac{1}{10}\right) = \frac{\ln 10}{k} \simeq \frac{2.30}{k},$$

so T is inversely proportional to the proportionate flow rate k.

4 Interpret the results

Collect relevant data

The only data required are the values of k. These can be determined for each lake from its volume V and water flow rate r, since $k = r/V$. The data are given in Table 1.1 for all of the Great Lakes.

Table 1.1

Lake	Volume V (10^{12} m^3)	Water flow rate r (10^3 m^3 s^{-1})	$k = r/V$ (10^{-9} s^{-1})
Superior	12.10	2.01	0.166
Michigan	4.92	1.57	0.319
Huron	3.54	5.10	1.441
Erie	0.48	5.90	12.292
Ontario	1.64	6.78	4.134

The units given mean, for example, that the volume of Lake Superior is $12.10 \times 10^{12}\,\text{m}^3$, its water flow rate is $2.01 \times 10^3\,\text{m}^3\,\text{s}^{-1}$, and its proportionate flow rate is $0.166 \times 10^{-9}\,\text{s}^{-1}$.

Describe the mathematical solution in words

For Lake Superior, for example, the model predicts that a reduction in the level of pollution to a tenth of its initial level will take

$$T = \frac{\ln 10}{k} = 13.9 \times 10^9 \text{ seconds.}$$

Dividing T by $60 \times 60 \times 24 \times 365.24$ to convert it into years, the model predicts that it will take 439 years for the pollution level in Lake Superior to drop by a factor of ten. A similar calculation for Lake Michigan predicts that it will take 7.22×10^9 seconds, or 229 years, for the pollution level to drop by a factor of ten.

According to Assumption (g) on page 7, only clean water enters the downstream lakes Huron, Erie and Ontario from the upstream lakes, so the model can be used for these lakes as well. Similar calculations then predict that the times required to reduce the pollution level to a tenth

It may be necessary to revisit this assumption later.

of its initial level are 51 years for Lake Huron, 6 years for Lake Erie and 18 years for Lake Ontario.

The reason for the long time periods that are predicted for purging the two upper lakes is their rather low proportionate flow rates. The quantity $V/r = 1/k$ is the *average retention time* for water in the lake, that is, the length of time it takes for an amount of water equal to the volume of water in the lake to flow out of the lake, and hence the time any molecule of water would expect to remain in the lake. For Lake Superior, the average retention time is 191 years, and that for Lake Michigan is 99 years. For the downstream lakes the average retention times are much shorter.

Decide what results to compare with reality

It is perhaps fortunate that the two most polluted lakes in the 1960s were Lake Erie and Lake Ontario, which have relatively small volumes of water and fairly high water flow rates. It was thus possible, as predicted by the model, to clean up these lakes fairly quickly.

You will see a model of the cumulative effects of pollution levels on the whole system in Subsection 6.2.

The model predicts that if significant amounts of pollution are allowed to enter Lake Superior or Lake Michigan, then it will take hundreds of years for the lakes to recover. It could be argued that the mathematical model is quite conservative in its estimation, since it assumes that all sources of pollution have stopped and that the pollutant is completely dispersed in the water, rather than in the flora and fauna or in the sediment at the bottom of the lake. However, since most pollutants biodegrade, it could also be argued that the estimate is rather pessimistic. Even so, a major contamination of either of these lakes, particularly by a pollutant which does not biodegrade, would be a catastrophe for which there would, apparently, be no short-term solution.

It is salutary to note that, in recent years, there have been Fish Advisory Warnings in place for many areas, and in particular for Lake Michigan. People are advised not to eat fish from that lake because of the high level of toxic chemicals in them. Although these chemicals are present in small quantities in the lakes, they tend to concentrate when moving up through the food chain, here leading to fish which accumulate significant quantities of pollutant chemicals. The battle to control pollution in these lakes has not yet been won.

This concludes the summary of the first four modelling stages for this problem. The fifth stage, 'Evaluate the model', will be addressed in Subsection 5.1.

The development of the pollution model for the Great Lakes has highlighted a number of important aspects of mathematical modelling. These include the following.

- By representing quantities using symbols, it is possible to develop models that can be applied in a variety of different situations.
- The simplifying assumptions must be relevant to the model, and should be linked to the mathematical relationships between the variables and parameters.
- Data for the model should be collected after it has been established exactly what data are required.

End-of-section Exercises

Exercise 1.1

In your own words, and without using any equations or symbols, give an outline, or description, of the formulation of the model in this section.

Exercise 1.2

(a) Suppose that a pollutant continues to enter a lake at a constant rate of $q\,\text{kg}\,\text{s}^{-1}$ and assume that this pollutant instantaneously diffuses throughout the lake. How would the mathematical model given by Equation (1.1) change?

(b) Show that the mass of pollutant in the lake is now modelled by

$$m(t) = \left(m(0) - \frac{q}{k}\right) e^{-kt} + \frac{q}{k}.$$

What is the long-term effect on the mass of pollutant in the lake?

(c) What happens if, initially, there is no pollution in the lake?

(d) What is the corresponding mathematical model for pollutant concentration?

2 Analysing a model

The Introduction described the stages of the mathematical modelling process, and in Section 1 you saw how some of those stages can be applied in the context of modelling pollution in the Great Lakes. In this section you are asked to relate the stages of the mathematical modelling process to a previously formulated mathematical model. This model was not developed using the structure for modelling described in the Introduction, and consequently some of the modelling stages have not been made explicit. This example is typical of accounts of modelling that you may see in books, or produced in the workplace. The aim of this section is to help you to draw out and to clarify mathematical modelling ideas by considering the example.

2.1 Skid marks

The description in the box overleaf includes all of the elements of mathematical modelling, but you may have to search a little to find them. The basic mathematical idea used is an equation for motion in one dimension with constant acceleration.

This is covered in MST209 *Unit 6*, but in slightly different notation.

First read the whole description of the mathematical model. Then turn to the extended exercise following the box. This consists of a number of questions about the model, drawing out the main modelling points. You will need to refer back to the modelling description as you work through the exercise.

As you read through the description of the model, bear in mind the points made in the previous section. Look out for the purpose of the model, the system that is being modelled, the simplifying assumptions, definitions of the variables and the derivation of relations between them, and the conclusions that are drawn from the model.

Modelling example: skid marks

When the police are investigating a road accident, the skid marks of a car can be very informative. From the length of the skid marks of the wheels, the police can estimate the speed at which the car was travelling before the wheels locked and the car went into the skid.

To assist in this estimation, the police sometimes drive a similar 'test' car with similar tyres and under similar road conditions, and cause it to skid at the same place, but at a lower speed. They then compare the skid marks produced by the test car with the original ones. They assume that the frictional forces between the surfaces are not dependent upon the speed of the car, only upon the mass of the car, the condition of the road and the type of surface. The frictional force is assumed to be proportional to the mass of the car, as are any accelerating or decelerating forces due to gravity if the car is going up or down a hill. The decelerations of the original car and the test car are the same according to these assumptions.

So it is necessary to use the test car's results to calculate the deceleration and then to use the calculated deceleration to estimate the speed of the original car.

For the test car, let us call the initial speed u_{test} and the distance x_{test}. The final speed v is 0, so

$$u = u_{\text{test}},$$
$$v = 0,$$
$$x = x_{\text{test}},$$
a_{test} is to be found.

The equation linking u, v, x and constant acceleration a is

$$v^2 = u^2 + 2ax, \qquad (2.1)$$

This equation is derived in MST209 *Unit 6* as
$$v^2 = v_0^2 + 2a_0x,$$
where v_0 is the initial velocity and a_0 is the constant acceleration. The equation is numbered here for reference in the following exercise.

so that

$$0 = u_{\text{test}}^2 + 2a_{\text{test}}x_{\text{test}},$$

i.e. $$a_{\text{test}} = \frac{-u_{\text{test}}^2}{2x_{\text{test}}}.$$

For the original car, if it skidded to rest over a distance x_{car}, assuming that its deceleration a_{car} is the same as that for the test car a_{test}, then

$$a = a_{\text{car}} = a_{\text{test}} = \frac{-u_{\text{test}}^2}{2x_{\text{test}}},$$
$$v = 0,$$
$$x = x_{\text{car}},$$
$u = u_{\text{car}}$ is to be found.

Using the constant-acceleration equation again gives

$$0 = u_{\text{car}}^2 - \frac{2u_{\text{test}}^2 x_{\text{car}}}{2x_{\text{test}}},$$

i.e. $$u_{\text{car}} = u_{\text{test}}\sqrt{\frac{x_{\text{car}}}{x_{\text{test}}}}.$$

This formula will be put to use in a computer activity in Subsection 6.1.

So the speed of the car before the accident can be estimated. Bear in mind that the speed of the original car would be greater than u_{car} if it had crashed, rather than stopped, at the end of the skid.

Exercise 2.1

(a) State in your own words the purpose of this model, and say when it may be useful.

(b) What is the role of the test car? Why could the police not just be issued with tables giving the speed in terms of the length of the skid marks?

(c) The basic concern of mathematical modelling is with finding relationships between variables that specify the system under consideration; in other words, to find formulae that enable you to calculate something if you know the value of something else, or that tell you how something varies with something else. What are the appropriate 'something' and 'something else' in this model (in words)? (The second sentence of the description of the model may be of help here.) On the basis of what you know about skids, say what you can, in the simplest terms, about the nature of the variation.

(d) Several symbols appear in the description of the model, but according to part (c) it is the relationship between just two key variables which is of concern. The symbols conveniently form four groups: symbols for the two key variables; symbols for the data; symbols introduced to make the calculations easier; and symbols used in a general formula employed in the model. Classify the symbols according to this scheme.

(e) No units of measurement are given anywhere in the definitions of the symbols. This is not the practice adopted in this course, and you are usually advised not to follow it. But does it matter in this instance?

(f) What is the basic model that underlies the whole discussion?

(g) The basic model identified in your answer to part (f) assumes particle motion with constant acceleration in a straight line. Are these assumptions mentioned in the example? What other assumptions are mentioned or are implicit in the example?

(h) Using Newton's second law and the properties of sliding friction, justify the assumption of constant acceleration (that is, constant deceleration for a skidding car), for the case where the road is flat.

You may find it helpful to draw a force diagram. Sliding friction is discussed in MST209 *Unit 6*.

(i) Justify the 'assumption' that the decelerations of the two cars are the same.

(j) The possibility that the accident took place on a slope is alluded to in the description of the model, but it is not stated explicitly whether the model applies to such a situation. Does it?

(k) Decide whether the model would apply to the following situations:

(i) a crashed Rolls-Royce;

(ii) a crash into a strong headwind;

(iii) a crash in a shower of rain, if the road dries out before the police arrive on the scene.

(l) The model, so far as the original car is concerned, might be summarized as follows.

'We want to determine the initial speed u_{car} in terms of the length of the skid x_{car} (which is the distance that the car travels before coming to rest). Under the assumption of constant deceleration in a straight line, this is given by Equation (2.1) as

$$u_{\text{car}}^2 + 2a_{\text{car}}x_{\text{car}} = 0, \qquad \text{or equivalently,} \qquad u_{\text{car}} = \sqrt{-2a_{\text{car}}x_{\text{car}}}.'$$

The car's acceleration, a_{car}, is negative because the car is decelerating, thus $-2a_{\text{car}}x_{\text{car}} > 0$.

Confirm that this result agrees with intuitive expectations, as given in the answer to part (c). Explain why this is not the end of the story, and how the test car comes in.

(m) By combining the formulae for the original car and for the test car, the final formula

$$u_{\text{car}} = u_{\text{test}}\sqrt{\frac{x_{\text{car}}}{x_{\text{test}}}}$$

is obtained. From this formula, the example says, 'the speed of the car before the accident can be estimated'. If the actual skid marks are four times as long as those of the test car, and the test car was going at 40 mph when it skidded, was the original car breaking the 70 mph speed limit?

(n) Suppose that you are a police instructor: explain to a police officer, who is about to go out on an accident investigation for the first time, how to estimate the speed of the car before the accident.

(o) Explain the final sentence of the example: 'the speed of the original car would be greater than u_{car} if it had crashed, rather than stopped, at the end of the skid'.

2.2 Commentary

Reading and interpreting descriptions of mathematical models can be hard work: there is usually a lot of information to absorb, and it can be difficult to focus on what is really important. For example, in the Great Lakes model, the names and volumes of the lakes do not help our understanding of the mathematics (though they are of crucial importance to a geographer). The process of coming to terms with a mathematical model is one of digging away until you find what lies at the root of it all. In the Great Lakes model it is the differential equation $dm/dt = -km(t)$ that lies at the root, whereas in the skid marks model it is the formula relating initial and final velocity to distance, for motion with constant acceleration in a straight line.

See Equation (2.1).

One of the basic skills of mathematical modelling (as you will find when you come to construct models for yourself) is to formulate the fundamental equation or relationship that describes the process in which you are interested. There are several points in the two accounts seen so far that show how this can be done.

- It is important to pick appropriate variables and parameters, and to define them carefully.
- It is necessary to simplify matters in order to make progress. For example, it was recognized in creating the Great Lakes model that the problem of seasonal variations in water level could be postponed, if not ignored.
- It is good practice to record the assumptions that you make in deriving the model. This will provide a clear basis for any further development.
- It is important to collect relevant data, both to check the predictions of the model and to furnish the values of any parameters that are needed to apply the model to any particular situation.

End-of-section Exercises

Exercise 2.2

This exercise is based on the skid marks modelling report and subsequent commentary.

(a) Discuss the advantages and disadvantages of using compound symbols for variables, such as u_{test} and u_{car}, instead of single symbols such as u and U.

(b) Why is it useful to classify parameters and variables as distinct categories?

(c) Do you consider the following statement to be an assumption? 'The value of the coefficient of friction is not greater than 1.0.'

(d) Data are necessary both to provide the values of parameters and to validate the model. What data do you think will be required to provide values of the parameters, and what to validate the model?

(e) Do you consider the skid marks modelling report to be easy to follow?

3 The skills of modelling

Mathematical modelling involves many different skills. To be good at mathematical modelling you have to be able to do some mathematics, but you need to be capable of doing other things as well. If you tend to think of mathematics as a set of procedures (such as the procedures for solving differential equations) then you may not regard these additional skills as being relevant to mathematics. Most practising mathematicians, concerned with pure as well as applied mathematics, would disagree with that interpretation. One of the most powerful motivations for studying mathematics is the desire to solve *new* problems for which there is no known solution procedure. The skills required in mathematical modelling include many general problem-solving skills. To be able to deal with mathematical modelling problems is more generally useful, and more difficult, than (say) being able to solve first-order differential equations by the integrating factor method. It calls upon skills of creativity, analysis and interpretation which apply to all sorts of problems, not just mathematical ones.

Here is a list of skills that may be required in the solution of a modelling problem, placed in the order of the modelling framework introduced earlier. You need to be able to do the following.

- Specify the purpose of the model, by defining or interpreting the problem you are investigating.
- Create the model by
  - simplifying the problem (by means of appropriate assumptions),
  - choosing appropriate variables and parameters,
  - formulating relationships between the variables.
- Use mathematics to find a solution from the relationships.
- Interpret the results by describing them in words (or otherwise) so that they can be understood by a possible user.
- Evaluate the model by
  - checking that the mathematical relationships and the solution make sense,
  - comparing the results with reality,
  - checking their sensitivity to changes in the data.

For the problem to be considered shortly, the 'Do mathematics' stage of modelling is very brief and is therefore included in the subsection for 'Create the model'.

When tackling a modelling problem in earnest, you have to call on these skills repeatedly, in a complicated and interactive way. It seems wise, therefore, to practise the skills individually at first, and this is the aim of the current section. Each subsection deals with a subgroup of skills from the list above, and is illustrated by reference to the two models considered in Sections 1 and 2. To try out these individual skills you are also asked to apply them to the following modelling problem. You are not left entirely on your own to tackle this problem. It is broken down later into steps corresponding to the modelling skills listed above.

The tin can problem

What shape should a tin can be? Most cans are cylindrical, so suppose that the best tin can is cylindrical. What shape of cylinder is best: short and fat, long and thin, or somewhere in between? Think of the standard kind of can in which soup, beans and other foodstuffs are preserved. Many of the cans in supermarkets contain about 400 grams of food and are rather taller than they are wide. There is a degree of uniformity in the general shape of such cans, which is quite surprising when there are so many different brands and when they contain so many different things; imported cans seem to have a shape similar to those that are home-manufactured. Is this traditional, or is it because they all conform to some ideal shape; and if the shape is not ideal, could there be advantages in changing it?

Although cans are made from a variety of materials, they are usually called *tin cans* because they were first made from tinned steel.

3.1 Specify the purpose of the model

In mathematical modelling, problems are rarely posed in a way that can be translated directly into mathematical form. For example, from the description of the treatment of pollution in the account of the Great Lakes model, you might have thought that, in order to construct a mathematical model, you would need to know quite a lot about pollution. However, it was possible to produce a straightforward non-technical statement of the underlying problem that could serve as the basis for a model, namely:

> to investigate how the pollution level varies with time as clean water flows into the lake and polluted water flows out.

It is important to establish at the outset a clear statement of the purpose of a model. For example, the purpose of the Great Lakes model is to discover the time that it will take for the pollution level in a lake to reduce to a given proportion of its initial pollution level. Such a statement is typical of the approach that is required in order to start a modelling problem.

Although a clear statement of the problem is necessary, it may change as the model develops, and the final statement may be different from that conceived at the outset. However, it is important to have a target at which to aim, even if this target changes during the process.

It is worth bearing in mind that some models, created for a specific purpose, may be applicable in other situations. It may also be possible to save time by modifying a model that has been used for one situation so that it can be applied to another. For example, the mathematical model developed to predict how long it would take for pollution in the Great Lakes to reduce to a target level could be adapted for use in drug therapy, or for the cleaning of milk churns by running water through them.

Exercise 3.1

Consider the problem of finding the best shape for a cylindrical tin can which is to contain a specified quantity of baked beans.

(a) The idea of 'best' occurs frequently in mathematical modelling, and its meaning needs to be made precise. What should the word 'best' mean in the phrase 'the best shape for a cylindrical tin can'?

(b) Try to formulate a clear statement of a suitable modelling problem, based on your answer to part (a).

There is no unique correct answer in modelling; if you reach a different answer to the one given, that does not mean that your answer is wrong. However, for the purposes of moving the story forward at each stage, the model will be developed in the text from the solution given to the preceding exercise.

3.2 Create the model

There are a number of skills that are needed when building a sensible model that approximates the real situation. In creating a model, these skills may be required at a variety of stages and not necessarily in the order in which they are presented here.

Simplify the problem by stating assumptions

The skid marks model depended on the results that the deceleration of a skidding car is constant and that, for given conditions, different cars have the same deceleration while skidding. These follow from assumptions that underpin a well-established theory of sliding friction, but hold only if (for example) air resistance is ignored. To ignore air resistance is justified on two counts: firstly, its effects are probably small compared with those of sliding friction; secondly, the resulting model is relatively easy to analyse, and may provide some insight into the problem. In modelling you should always look for as simple a model as possible, consistent with the principal features of the problem. (To have ignored the effects of friction would obviously have been counter-productive.) It is important to be clear about the simplifying assumptions that have been made in order to arrive at the model. Recording an explicit list of the assumptions makes it easier for the reader to follow the development of the model, and should you need to improve your model, you then have an obvious place to start: review the assumptions, and ask which should be modified or relaxed.

In later sections of this text, you will see how to check whether simplifying assumptions are justifiable.

Exercise 3.2

Continuing with the tin can problem, as specified in Exercise 3.1(b), what simplifying assumptions, if any, need to be made?

Choose appropriate variables and parameters

Identification of the key variables is of paramount importance. If you can summarize the problem in terms of describing roughly how one quantity varies with another, or several others, then you should have no difficulty in identifying the key variables. Once these key variables have been identified, it should be possible to obtain relationships between them, which may throw up other variables and/or parameters. It is good practice to keep a list of all the variables and parameters, adding to it as necessary, to ensure that all of them have been consistently defined and used. It pays to be careful in defining variables and parameters, to avoid confusion later. For example, 'time since all pollution ceased' is clearer, and less likely to be misinterpreted, than just 'time'.

In the case of the Great Lakes model, there are three key variables: the first is the time (in seconds) since all sources of pollution ceased, the second is the mass (in kilograms) of pollutant in the lake, and the third is the concentration (in kilograms per cubic metre) of pollutant in the lake. The creation of the mathematical model involves identifying relationships between these key variables. In writing down the relationships, five parameters were identified: the target concentration level (in kilograms per cubic metre) of pollutant in the lake; the time taken (in seconds) to reduce the concentration of pollutant to the target level; the volume (in cubic metres) of water in the lake; the water flow rate (in cubic metres per second); and the proportionate flow rate (in seconds^{-1}).

In the skid marks model, there are two key variables: the length of the skid for the original car, and its initial speed, each in appropriate units. In writing down the relationship between these two key variables, two further variables and a parameter were identified: the length of the skid for the test car, its initial speed, and the common deceleration of the two cars. The units for these quantities need to match the units chosen for the key variables. (The final speeds of the two cars could be regarded as additional parameters.)

Exercise 3.3

Continuing with the tin can problem, define the variables and parameters that you think will be needed, giving appropriate units.

Formulate relationships

The use of the input–output principle to formulate relationships in creating the Great Lakes model is a good illustration of quite a common modelling technique. Another common technique for formulating relationships, in the case of mechanics problems, is to make use of Newton's laws of motion, although their use may be hidden, as in the skid marks model.

Often it is helpful to draw a diagram. Not only does a diagram help in the definition of the variables and parameters, but it tends to help in gathering together some key factors.

A picture is said to be worth a thousand words.

Exercise 3.4

Continuing with the tin can problem, draw a diagram to help with the definition of the variables and parameters.

Exercise 3.5

Write down formulae that relate the variables. You should explain on which assumptions any formula is based. Derive a formula that relates the area A of the can to its radius r, where the volume V is a parameter.

Exercise 3.6

Are there any assumptions that have not been used in the formulation? Are they needed?

Exercise 3.7

Are there any variables or parameters not used in the derivation?

Find a solution

Once the formulae that relate variables have been derived, some mathematics will probably be needed to find a solution to the model. In the skid marks model, the solution for u_{car} of the equation

$$0 = u_{\text{car}}^2 - \frac{2u_{\text{test}}^2 x_{\text{car}}}{2x_{\text{test}}}$$

was required. In the Great Lakes model a differential equation was solved, an initial condition was used, and an algebraic equation had to be solved to find the target time.

Typical mathematical techniques used in simple models are: solving algebraic equations; solving differential equations; finding an optimum value.

Exercise 3.8

Continuing with the tin can problem, use the final formula from Exercise 3.5 to find 'the best shape for a cylindrical tin can'.

3.3 Interpret the results, evaluate the model

Interpret results

Evaluate

Obtaining a mathematical solution to a modelling problem is not the end of the modelling process. The solution needs to be interpreted in terms of the original problem posed and a number of checks should be made. This subsection outlines some of the techniques used to interpret the solution and to check its reliability.

Check that the model and solution make sense

It is sometimes possible to take some particular values for the variables, and so make a quick check on the correctness of the model. In the skid marks example, it makes sense that the longer the skid mark of the car, the greater the speed that it was travelling beforehand, and this is borne out by the solution. Also, if the length of the skid mark is zero, then the model and intuition both give the same value, zero, for the speed of the car. In the Great Lakes model, the time taken to reach the target pollution level is increased if the target level is decreased, and this makes sense. It is always worth investigating the solution with checks such as these in mind.

Exercise 3.9

(a) How would you expect the area of the tin can to change as the radius becomes very small? Is this what the model predicts?

(b) How would you expect the area of the tin can to change as the radius becomes very large? Is this what the model predicts?

(c) Is there any other test that can be applied to check whether the solution is reasonable?

Compare the results with reality

A check that the model predicts the kind of results that one would expect from common sense and from experience, as described above, is one type of comparison with reality. Beyond that, if possible, one should validate the model by comparing its predictions with data from an experiment or other reliable source. It is good practice to try to reformulate the results so that this check turns into something simple such as drawing a straight line on a graph. You may also require some data to give an explicit numerical solution to the problem, such as the values of the lake volume and water flow rate in the Great Lakes problem. The practicality of the skid marks model lies in the use of the test car to provide these data.

You will see a check that involves 'drawing a straight line' in Section 4.

Exercise 3.10

(a) Continuing with the tin can problem, summarize the solution obtained to the problem descriptively, giving the optimal shape of the can in terms of the ratio of height to radius.

(b) Measure the radius and volume of some tin cans. Does the shape of can predicted by your solution correspond to the actual shape of tin cans?

The volume can be taken either as the stated volume of the contents or as an estimate based on measurements of the diameter and height.

Exercise 3.11

(a) Consider the assumptions that were made in Exercise 3.2, and decide which of these might be the main source of the discrepancy between the solution to the model and reality.

(b) Relax the assumption of no wastage, by assuming now that each circular end for a can is cut from a square of tin plate of side $2r$. Modify the formula that was obtained in Exercise 3.5 to take account of this change. Solve the modified problem, and comment on your answer.

Exercise 3.12

In this exercise you are asked to reconsider the underlying assumption that the cost of making the tin can is proportional to the area of tin plate used. Suppose instead that the soldering of joins is the most costly part of making cans. The length L of metal to be soldered on each can consists of the circumferences of the top and bottom circular pieces plus the height of the can. Derive an expression for L in terms of the radius r, where the volume V is a parameter. Hence determine the ratio of height to radius which will minimize the amount of soldering required.

Check the sensitivity of the solution

In many models the parameters are estimated, and it is useful to know whether the solution is sensitive to small changes in the value of a particular parameter. If the solution is insensitive to such small changes, then it is not worth spending time and trouble in finding a better estimate for the parameter, whereas if the solution is sensitive to changes in the value of a particular parameter, then it is necessary to obtain a very good estimate for this parameter.

At the residential school, there will be a more systematic approach to sensitivity analysis. In this text, an experimental approach is adopted.

For the tin can problem, the volume is only an estimate, obtained either from the stated volume of contents on the label or by measurement. In the next exercise, you are asked to find the effect on the optimum radius of small changes to the volume of the tin can.

Exercise 3.13

Consider the simple model for the tin can problem developed in Exercise 3.8.

(a) Suppose that the tin can has a nominal volume of 400 ml. What will be the radius of can for which the area is a minimum?

(b) Now consider a 1% change in the volume of the can. By what percentage does the optimal radius change? Do you think that the predicted optimal radius of the can is sensitive to changes in the value of the volume?

(c) What does this tell you about the accuracy to which you need to measure the volume of tin cans?

Note that

$$1 \text{ litre (l)} = 1000\,\text{cm}^3,$$

so that

$$1 \text{ millilitre (ml)} = 1\,\text{cm}^3.$$

End-of-section Exercises

Exercise 3.14

In your own words, and without using any equations or symbols, give an outline of the formulation of the first model.

---

4 Transient heat transfer: a case study

In the previous section, the idea of revising the model was introduced. In this section you will be taken through the whole modelling process in detail, from creating a first simple model, through evaluating it, to the subsequent revision of the model by changing one of the assumptions. The new aspect here is the emphasis on a revised model, which comes in Subsection 4.2. The problem that will be examined is one based on heat transfer.

This first model is also developed in MST209 *Unit 16* but the revision is not.

> ***The cooling of a cup of tea***
>
> Immediately after a cup of tea is made, it is usually too hot to drink. How long does it have to be left to cool before it is drinkable? More generally, how does the temperature of the cooling tea vary with time?

This is not, on the face of it, an important problem. However, there are many situations in which cooling plays a significant role, some of which are of considerable importance, particularly in industry. Analogous domestic situations include the cooling of a hot-water tank and the warming of a cool box. The problem of the cooling of a cup of tea has been chosen because it is a familiar and simple example, which involves the principles that apply in most cooling problems and for which a simple experiment can be used to test the conclusions of the model.

Heat is most commonly transferred by conduction, convection and radiation. The cooling of a cup of tea will occur through each of these three heat transfer modes. Also, when the tea is hot, steam rises from its surface, so it is likely that evaporation plays a role in the cooling process. However, to keep the modelling simple, consider initially just one form of heat loss: convection from the surface of the tea. As with all modelling, it is better to develop first a simple model, which you can be reasonably confident will lead to an answer.

4.1 The first model

Exercise 4.1

Make a list of the assumptions that you think should be made in deriving a first model for the cooling of a cup of tea. (This should be consistent with heat loss from the tea occurring only by convection from the surface of the tea.)

The various aspects of steady-state heat transfer referred to in this section are introduced in MST209 *Unit 15*.

The variables and parameters to be used in developing the model are listed in the following table.

Table 4.1 Variables and parameters for the first model

Symbol	Definition	Units
t	time since the tea started to cool	s
$\Theta(t)$	temperature of the tea at time t	°C
m	mass of the tea	kg
c	specific heat (capacity) of tea	$\mathrm{J\,kg^{-1}\,K^{-1}}$
δE	change in the heat energy of the tea during a small time interval $[t, t+\delta t]$	J
Θ_{air}	ambient temperature of the surrounding air	°C
$\Theta_{\text{sur}}(t)$	temperature at the surface of the tea at time t	°C
$q(t)$	rate of heat transfer from the tea to the air, through the surface of the tea, at time t	W
D	inside diameter of the cup at its rim	m
A	surface area of the tea	$\mathrm{m^2}$
h_{tea}	convective heat transfer coefficient between the tea and the surface	$\mathrm{W\,m^{-2}\,K^{-1}}$
h_{air}	convective heat transfer coefficient between the surface and the air	$\mathrm{W\,m^{-2}\,K^{-1}}$
U	U-value for the surface of the tea	$\mathrm{W\,m^{-2}\,K^{-1}}$
Θ_0	initial temperature of the tea	°C
Θ_T	temperature at which the tea is drinkable	°C
T	time required for the tea to become drinkable	s

Exercise 4.2

Which of the symbols in Table 4.1 are variables?

Exercise 4.3

Some of the symbols chosen have subscripts, for example, $\Theta_{\text{sur}}(t)$. Would it be more convenient to replace these subscripted variables by single-letter symbols?

The simplest part of the formulation is to relate the surface area of the tea to the inside diameter of the tea cup, and by Assumption (i), this is

$$A = \pi\left(\tfrac{1}{2}D\right)^2.$$

The assumptions are listed in the solution to Exercise 4.1, on page 55.

The approach that is adopted here is to consider the changes that occur over a small time interval from t to $t + \delta t$. Heat is lost from the tea during this interval, and the input–output principle is applied to describe the heat exchange between the tea and its surroundings.

Because of Assumptions (b) and (c), only convective heat transfer need be considered. The standard model for steady-state convective heat transfer from a surface of area A to an adjacent fluid, is given by the formula

$$q = hA(\Theta_1 - \Theta_2), \tag{4.1}$$

See MST209 *Unit 15*. Assumption (d) says that this model may be applied here.

where h is the convective heat transfer coefficient, and the temperatures on the surface and in the body of the fluid are Θ_1 and Θ_2. There are two convective heat transfers at the surface of the tea: one from the tea itself to the surface, and one from the surface to the air. By Assumptions (d) and (e), the convective heat transfer from the tea to the surface is modelled by

$$q = h_{\text{tea}}A(\Theta - \Theta_{\text{sur}}), \tag{4.2}$$

Here $\Theta_1 = \Theta$ and $\Theta_2 = \Theta_{\text{sur}}$.

where q is the rate of loss of heat energy from the cup of tea, h_{tea} is the convective heat transfer coefficient between the tea and the surface, and A

is the area of the surface of the tea. Similarly, by Assumptions (d) and (f), the convective heat transfer from the surface to the air is modelled by

$$q = h_{air}A(\Theta_{sur} - \Theta_{air}), \qquad (4.3)$$

Here $\Theta_1 = \Theta_{sur}$ and $\Theta_2 = \Theta_{air}$.

where h_{air} is the convective heat transfer coefficient between the surface and the air, and Θ_{air} is the ambient temperature of the air.

Exercise 4.4

Eliminate Θ_{sur} from Equations (4.2) and (4.3) to obtain a single equation modelling the rate of convective heat transfer from the tea to the air.

In the solution to Exercise 4.4 you found that the heat loss from the tea to the air is given by

$$q = UA(\Theta - \Theta_{air}), \qquad (4.4)$$

where

$$U = \left(\frac{1}{h_{tea}} + \frac{1}{h_{air}}\right)^{-1}. \qquad (4.5)$$

The equation $q = UA(\theta - \theta_0)$ occurs frequently in heat transfer models. The number, and arrangement, of the conducting and convecting layers determine the value of U, which conveniently hides complications.

Now Equation (4.1) applies when the temperature differences remain constant in time. This is definitely not true of the situation under consideration; indeed, the whole point of the exercise is to find how the temperature of the tea *varies* with time. However, Assumption (g) says that the steady-state model will apply approximately over a short time interval $[t, t + \delta t]$, so that within this interval the rate of loss of heat energy by convection from the surface of the tea can be approximated by

This type of assumption is a standard modelling device.

$$q(t) \simeq UA(\Theta(t) - \Theta_{air}).$$

Note that q and Θ are now considered to be functions of the time t.

The quantity of heat energy lost over the time interval can therefore be approximated by

$$q(t)\delta t \simeq UA(\Theta(t) - \Theta_{air})\delta t,$$

and this is the *output* of heat energy from the tea over the interval. At the beginning of the time interval the temperature of the tea is $\Theta(t)$, whereas at the end of the interval it is $\Theta(t + \delta t)$. By Assumptions (a) and (h), the *accumulation* δE of heat energy in the tea over the interval $[t, t + \delta t]$ is

$$\delta E = mc(\Theta(t + \delta t) - \Theta(t)).$$

The constant c is the specific heat (capacity) of the tea; see MST209 *Unit 15*.

Since there is no input of heat energy to the tea during the interval, the *input* is zero. Applying the input–output principle,

$$\boxed{\text{accumulation}} = \boxed{\text{input}} - \boxed{\text{output}},$$

to the amount of heat energy in the tea over the time interval $[t, t + \delta t]$, now gives

$$mc(\Theta(t + \delta t) - \Theta(t)) \simeq 0 - UA(\Theta(t) - \Theta_{air})\delta t.$$

Rearranging this equation gives

$$\frac{\Theta(t + \delta t) - \Theta(t)}{\delta t} \simeq -\frac{UA}{mc}(\Theta(t) - \Theta_{air}). \qquad (4.6)$$

By Assumption (g), this approximation becomes more and more accurate as δt becomes smaller, and in the limit as $\delta t \to 0$, Equation (4.6) becomes the differential equation

$$\frac{d\Theta}{dt} = -\lambda(\Theta(t) - \Theta_{air}), \qquad (4.7)$$

This differential equation is of a form studied in MST209 *Unit 2*. The introduction of λ here is another example of reparametrization.

where $\lambda = UA/(mc)$. This model of the cooling process is usually referred to as **Newton's law of cooling**.

This completes the second stage (create the model) of the modelling process outlined in Section 1; assumptions have been stated, variables and parameters chosen, and mathematical relationships formulated.

Exercise 4.5

Have all the assumptions been used in the formulation?

The aim of finding how the temperature Θ of the tea changes with time t can be achieved by solving Equation (4.7). But before doing so, it is worth asking whether this equation predicts the right kind of behaviour for Θ.

This can be regarded as part of evaluating the model, a later stage of the modelling process. However, it is a natural check to carry out at this stage.

Exercise 4.6

(a) Qualitatively, how would you expect the rate of change of temperature $d\Theta/dt$ to depend on the temperature Θ of the tea?

(b) Is Equation (4.7) consistent with what you expect?

Sometimes Θ is written instead of $\Theta(t)$, when it is clear from the context that it is a function.

Having confirmed that the differential equation predicts the right kind of behaviour for Θ, the next task is to solve it, taking Θ_0 to be the initial temperature of the tea.

Do mathematics

Exercise 4.7

Find the particular solution of the differential equation

$$\frac{d\Theta}{dt} = -\lambda(\Theta - \Theta_{air})$$

that satisfies the initial condition $\Theta(0) = \Theta_0$.

So the required particular solution of the differential equation is

$$\Theta(t) = \Theta_{air} + (\Theta_0 - \Theta_{air})e^{-\lambda t}, \quad \text{where} \quad \lambda = \frac{UA}{mc}. \tag{4.8}$$

Does this formula agree with common sense? For $\Theta_0 > \Theta_{air}$, Equation (4.8) says that the temperature Θ decreases with time and tends to Θ_{air} in the long term. The temperature difference $\Theta - \Theta_{air}$ decreases exponentially with time, and so decreases at a slower rate for smaller temperature differences. So far, there is agreement with what might be expected. The temperature is predicted to fall more quickly if the surface area A is increased: that seems right, because there will then be more surface area from which the heat energy can dissipate. On the other hand, if there is more tea (larger m), then it will cool more slowly: a greater mass holds more heat energy, and the loss of a given amount of heat energy will therefore result in a smaller drop in temperature.

Again, this is an early part of evaluating the model, but worth doing at this stage.

It remains to answer the original specific question: how long will it be before the tea is drinkable? This is answered simply by setting $\Theta(t)$ equal to the temperature at which tea becomes drinkable, Θ_T, and solving for t. Suppose that $\Theta = \Theta_T$ at time $t = T$, that is, $\Theta(T) = \Theta_T$. From Equation (4.8),

$$\Theta_T - \Theta_{air} = (\Theta_0 - \Theta_{air})e^{-\lambda T}, \quad \text{so that} \quad \frac{\Theta_T - \Theta_{air}}{\Theta_0 - \Theta_{air}} = e^{-\lambda T},$$

and hence

$$T = \frac{1}{\lambda}\ln\left(\frac{\Theta_0 - \Theta_{\text{air}}}{\Theta_T - \Theta_{\text{air}}}\right) = \frac{mc}{UA}\ln\left(\frac{\Theta_0 - \Theta_{\text{air}}}{\Theta_T - \Theta_{\text{air}}}\right). \tag{4.9}$$

In order to obtain an explicit numerical answer, the values of the various parameters in Equation (4.9) need to be known. Table 4.2 below gives the relevant convective heat transfer coefficients, and values for h_{air} and h_{tea} will be taken from the right-hand column of this table.

Table 4.2 Values of convective heat transfer coefficients

Process	Fluid	Range of values of convective heat transfer coefficient h ($\text{W m}^{-2}\text{K}^{-1}$)	Assumed value of convective heat transfer coefficient h ($\text{W m}^{-2}\text{K}^{-1}$)
Free convection	Gas	2–25	10
	Liquid	50–1000	500
Forced convection	Gas	25–250	150
	Liquid	50–20 000	1000

Source: Incropera, F.P. and de Witt, D.P. (1990) *Introduction to Heat Transfer*, Wiley.

Table 4.2 is reproduced from MST209 *Unit 15*.

Table 4.3 below shows values of the other parameters obtained for a cup of tea, where in each case an explanation is given of how each value was found.

Table 4.3 Other parameter values

Parameter	Value	How obtained
m	0.25 kg	Weigh a full and an empty cup.
c	$4190\,\text{J kg}^{-1}\text{K}^{-1}$	Use known value for water.
U	$9.804\,\text{W m}^{-2}\text{K}^{-1}$	Use Equation (4.5) and Table 4.2.
D	0.072 m	Measure inside diameter of cup at rim.
Θ_0	80 °C	Measure with thermometer.
Θ_{air}	17.5 °C	Measure with thermometer.
Θ_T	60 °C	Sample, then measure with thermometer.

The ambient temperature of the air varied from 18 °C to 17 °C during the cooling period, so the average 17.5 °C is used here.

With these values, $\lambda = 3.81 \times 10^{-5}\,\text{s}^{-1}$ and $T = 10\,120$ seconds (rounded to the nearest ten seconds). In other words, the time taken for the tea to cool enough to be drinkable is estimated to be about 2 hours 50 minutes.

How does the model compare with reality? A member of the course team measured the temperature of the tea in a cup at home, and found that the tea took considerably less time, namely 12 minutes 50 seconds, to cool from 80 °C to 60 °C. So the numerical estimate provided by the model does not compare well with reality. Nevertheless, the qualitative prediction of an exponential decrease of temperature, given by Equation (4.8), may compare well with reality, at least for a different value of λ. In Table 4.4 (opposite) the temperature of the tea from this experiment is tabulated over a period of 90 minutes. A plot of this data is shown in Figure 4.1. From this graph, it appears that the data might well be fitted by a decreasing exponential function.

In order to test this hypothesis more closely, rewrite Equation (4.8) as

$$\Theta(t) - \Theta_{\text{air}} = (\Theta_0 - \Theta_{\text{air}})e^{-\lambda t}$$

and take the logarithm of both sides, to obtain

$$\ln(\Theta(t) - \Theta_{\text{air}}) = \ln(\Theta_0 - \Theta_{\text{air}}) - \lambda t.$$

It follows from this that a plot of $\ln(\Theta(t) - \Theta_{\text{air}})$ against t should, according to the model, produce a straight-line graph. This plot is shown in Figure 4.2, from which it is clear that the data points do not lie precisely on a straight line. So the tea did not cool quite exponentially: either its initial rate of cooling was too great, or its final rate of cooling was too small, or both.

One would not expect the data points to lie exactly on a straight line, given measurement inaccuracies. However, the data plotted in Figure 4.2 look as if they would be better fitted by a curve.

Table 4.4 Experimental data

Time (minutes)	Temperature of tea (°C)
0	80
1	77
2	75
3	$73\frac{1}{2}$
4	72
5	$70\frac{1}{2}$
6	69
7	$67\frac{1}{2}$
8	66
9	65
10	$63\frac{1}{2}$
12	61
14	59
16	57
18	55
20	53
25	$49\frac{1}{2}$
30	46
35	$43\frac{1}{2}$
40	41
45	39
60	$34\frac{1}{2}$
75	$30\frac{1}{2}$
90	$27\frac{1}{2}$

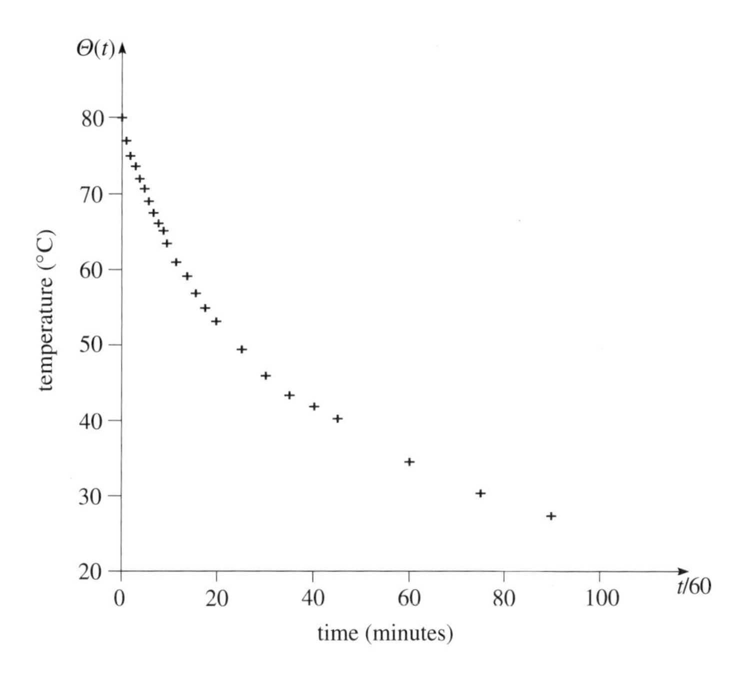

Figure 4.1 Graph of temperature against time for the cooling of tea in a cup

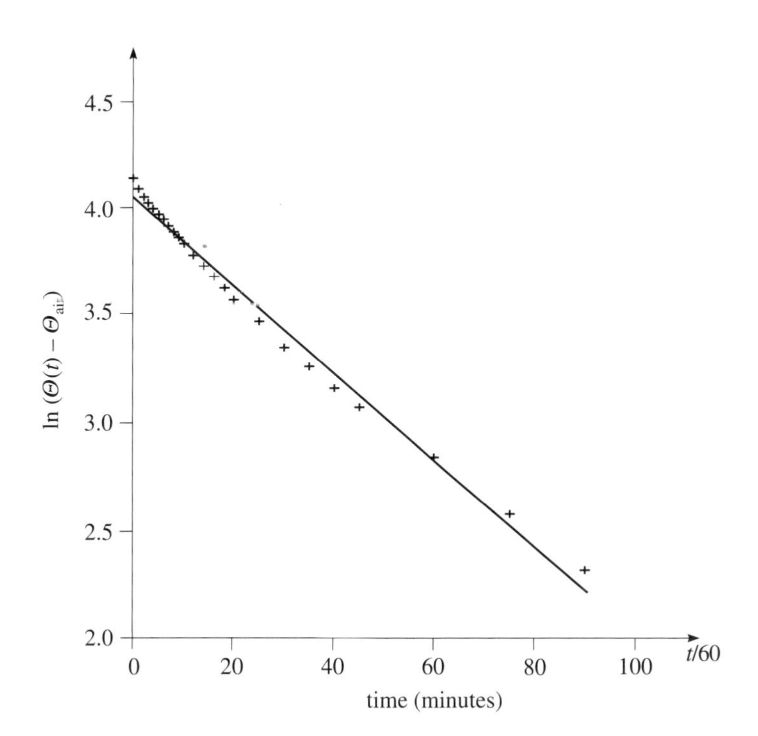

Figure 4.2 Graph of $\ln(\Theta(t) - \Theta_{\text{air}})$ against time and a straight-line fit to the data

Exercise 4.8

Ignoring for the moment that the graph in Figure 4.2 is not quite a straight line, and using a least-squares straight-line fit to the data points, the experimental value obtained for λ is $3.403 \times 10^{-4}\,\mathrm{s}^{-1}$, which differs from the value obtained using Table 4.3 by a factor of ten. Calculate the value of $U = (1/h_{\mathrm{tea}} + 1/h_{\mathrm{air}})^{-1}$ that would result in this value for λ, assuming that the other data values in Table 4.3 are correct. With $h_{\mathrm{tea}} = 500\ \mathrm{W\,m^{-2}\,K^{-1}}$, as before, what is the corresponding value of h_{air}?

Least-squares straight-line fits are discussed in MST209 *Unit 9*.

There is a discrepancy between the prediction from the model and the value from the experiment. In the solution to Exercise 4.8 you saw that, to predict the actual rate of cooling, a convective heat transfer coefficient from the surface to the air of approximately $106\,\mathrm{W\,m^{-2}\,K^{-1}}$ would be required, assuming that all the other data values are correct. This value is over ten times larger than the value given in Table 4.2, and is outside the range of typical free convective heat transfer coefficients for gases given there. This suggests that the discrepancy in the value of λ may not be due solely to uncertainties in the value of U.

This first simple model has some resemblance to reality; the model predicts an exponential decay, and the experiment shows a similar type of behaviour. However, the prediction for the time it takes the tea to cool to a drinkable temperature is in error by a factor of 10. The model needs to be revised, and this is attempted in the next subsection.

4.2 Revisions to the model

In considering how to revise the model, the starting point should be to consider how well the purpose of the model has been achieved. For the cooling of the cup of tea there were two objectives, neither of which was met satisfactorily when the model was tested using experimental data.

- The model predicts that the cup of tea should take about 2 hours and 50 minutes to cool down from 80 °C to 60 °C whereas, experimentally, this time was measured as 12 minutes 50 seconds.
- The model predicts an exponential relationship between the temperature of the tea and the time since the tea started to cool. The data seem to suggest that this is not precisely the case; it is clear that the plot of $\ln(\Theta(t) - \Theta_{\mathrm{air}})$ against t is not quite a straight line.

One possible reason for the first of these discrepancies is that the value used for $\lambda = UA/(mc)$ may not be very accurate. This is so because U depends on the values of h_{tea} and h_{air}, which are both hard to estimate with any precision. The values quoted in Table 4.2 for these two parameters are chosen from ranges of possible values, but in reality h_{tea} could take any value in the range 50–1000 and h_{air} any value in the range 2–25, giving values for U in the approximate range 2–24.

Exercise 4.9

The U-value is given in Equation (4.5) by $U = (1/h_{\text{tea}} + 1/h_{\text{air}})^{-1}$.

(a) Calculate the value of U, with $h_{\text{air}} = 10$ and h_{tea} taking, in turn, the values 50 and 1000 (the extreme values of the possible range for h_{tea}).

(b) Calculate the value of U, with $h_{\text{tea}} = 500$ and h_{air} taking, in turn, the values 2 and 25 (the extreme values of the possible range for h_{air}).

(c) What do you conclude from the previous two calculations about the sensitivity of U to changes in h_{air} and h_{tea}?

Exercise 4.9 shows that the value of U is dominated by the value of h_{air}, so the value of the cooling time obtained from the model will be more influenced by a change to the estimated value for h_{air} than by a similar change to h_{tea}.

Exercise 4.10

Explain why obtaining a better estimate for h_{air} would not satisfactorily address either of the deficiencies of the model described above.

So, although it would be sensible to try to obtain a better estimate for h_{air}, such a change addresses neither deficiency satisfactorily. Other ways of revising the model will need to be considered. Two possible revisions are presented in the remainder of this subsection.

When revising a model, it is better to try one revision at a time, so that the effects of each revision can be seen clearly; thus here each revision is done separately.

First revision: heat transfer through the sides

A revision should change the model so that it predicts a higher rate of heat loss. Assumption (c) said that the heat loss from the sides and bottom of the cup was negligible. In this first revision to the model, the rate of heat loss through the sides of the cup as well as from the surface of the tea is considered. The effect of such a revision on the model will be in the right direction: the prediction for the time taken for the tea to cool will be less.

To include heat losses by conduction and convection through the sides of the cup, some of the assumptions need to be revised and some new ones need to be made.

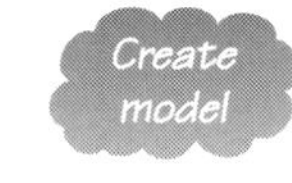

Exercise 4.11

Suggest a suitable simplifying assumption for the shape of the cup.

Assume that the cup is cylindrical (so it is more like a mug than a cup) and then use the standard model for steady-state heat transfer by conduction and convection through a uniform cylinder. However, since the thickness of the wall of a cup is small, the surface areas of the inside and outside of the cylindrical cup will be approximately the same. By making the assumption that the difference between these surface areas is negligible, the cylindrical part of the cup can be thought of as being formed from a thin slab folded round on itself and joined together, as shown in Figure 4.3 (overleaf). Thus the simpler model for a uniform slab can be used to model the heat transfer through the cylindrical surface.

This model, and that for a uniform slab referred to below, are developed in MST209 *Unit 15*.

Looking at the list of assumptions (in the solution to Exercise 4.1, on page 55), it is clear that Assumptions (c), (e), (f) and (g) need to be revised, and new assumptions need to be added. The full set of assumptions (with changes highlighted in italics and omissions marked by ...) is as follows:

(a) the change in heat energy of the tea is proportional to the change in its temperature and to the mass of the tea;

(b) all heat losses due to radiation are ignored;

(c) all heat losses due to conduction and convection through the ... bottom of the cup are ignored;

(d) in the steady state, the rate of heat transfer by convection at a surface between a fluid and another substance is proportional to the difference between the temperature of the surface and the temperature of the substance, and proportional to the area of the surface;

(e) the tea, apart from a thin layer close to its surface with the air *and to the sides of the cup*, has a uniform temperature;

(f) the surrounding air, apart from a thin layer close to the surface of the tea *and to the sides of the cup*, has a uniform constant temperature;

(g) over a short time interval, the steady-state model for the rate of heat transfer by *conduction and* convection provides a good approximation to the rate of heat energy loss, and this approximation improves as the time interval becomes shorter;

(h) changes in the heat energy of the material of the cup are ignored;

(i) *the cup is cylindrical in shape*;

(j) *the thickness of the cup wall is small when compared to other dimensions, so that the internal and external surface areas of the cup can be taken to be equal*;

(k) *in the steady state, the rate of heat transfer by conduction through a uniform slab is proportional to the difference between the temperatures of the two surfaces of the slab, proportional to the surface area of the slab, and inversely proportional to the thickness of the slab.*

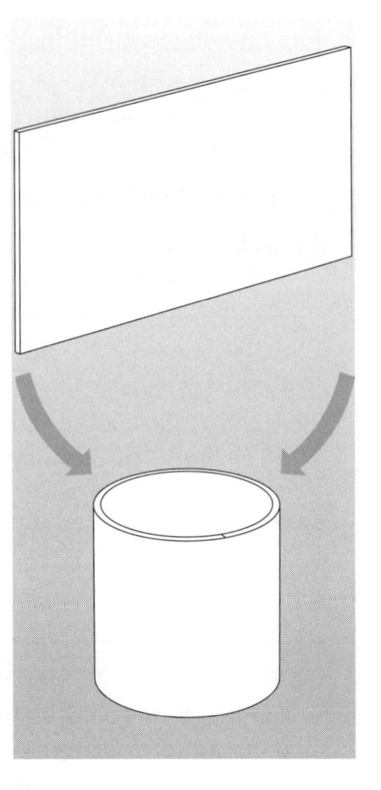

Figure 4.3

To develop the model, some new variables and parameters are introduced. Assume that h_{tea} and h_{air}, the convective heat transfer coefficients for the surface of the tea, also apply at the internal and external surfaces of the cup, respectively, then the additional variables and parameters required are as shown in Table 4.5.

Table 4.5 Additional variables and parameters for the first revised model

Symbol	Definition	Units
$q_{\text{side}}(t)$	rate of heat transfer through the sides of the cup at time t	W
H	height of cup	m
A_{side}	surface area of the sides of the cup	m^2
U_{side}	U-value for the sides of the cup	$\text{W m}^{-2}\,\text{K}^{-1}$
b	thickness of the wall of the cup	m
κ	thermal conductivity of the cup	$\text{W m}^{-1}\,\text{K}^{-1}$
$q_{\text{total}}(t)$	total rate of heat transfer through sides of cup and via convection at surface of tea at time t	W

Since h is being used for the convective heat transfer coefficients, a different symbol is needed for the height here, to avoid confusion.

By Assumption (j), the area A_{side} of the sides is the same for the inside and outside of the cup. Using Assumption (i),

$$A_{\text{side}} = \pi D H.$$

Assumption (j) models the wall of the cup as a uniform slab. Hence, using Assumptions (d), (g) and (k), the model for heat transfer by conduction and convection through the sides of the cup becomes

$$q_{\text{side}}(t) = U_{\text{side}} A_{\text{side}} (\Theta(t) - \Theta_{\text{air}}), \quad \text{where } U_{\text{side}} = \left(\frac{1}{h_{\text{tea}}} + \frac{b}{\kappa} + \frac{1}{h_{\text{air}}} \right)^{-1}.$$

Combining this with Equation (4.4), the total rate of heat transfer is

$$\begin{aligned} q_{\text{total}}(t) &= q(t) + q_{\text{side}}(t) \\ &= UA(\Theta(t) - \Theta_{\text{air}}) + U_{\text{side}} A_{\text{side}} (\Theta(t) - \Theta_{\text{air}}) \\ &= (UA + U_{\text{side}} A_{\text{side}})(\Theta(t) - \Theta_{\text{air}}). \end{aligned}$$

Notice that the total rate of heat loss is still proportional to the overall temperature difference; only the factor UA has changed, to $UA + U_{\text{side}} A_{\text{side}}$. Thus the input–output principle leads to the same differential equation (4.7), but now with

$$\lambda = \frac{UA + U_{\text{side}} A_{\text{side}}}{mc}.$$

Hence Equations (4.8) and (4.9) also still apply, with this new value for λ.

Exercise 4.12

How does the revised model address the purposes of the model, which were to predict the time for the tea to cool to a drinkable temperature and to determine how the temperature of the cooling tea varies with time?

Interpret results

Values for the additional parameters are shown in Table 4.6.

Table 4.6 Values for the additional parameters of the revised model

Parameter	Value	How obtained
H	0.0615 m	Measure height of cup.
b	0.003 m	Measure thickness of cup wall.
κ	$1.5\,\text{W m}^{-1}\,\text{K}^{-1}$	From MST209 *Handbook*.

These data and those in Table 4.3 give

$$A_{\text{side}} = \pi \times 0.072 \times 0.0615 \simeq 0.0139$$

and

$$U_{\text{side}} = \left(\frac{1}{h_{\text{tea}}} + \frac{b}{\kappa} + \frac{1}{h_{\text{air}}} \right)^{-1} = \left(\frac{1}{500} + \frac{0.003}{1.5} + \frac{1}{10} \right)^{-1} \simeq 9.615,$$

Note that this U-value, like U previously, is dominated by the value of h_{air}.

so that

$$\lambda = \frac{UA + U_{\text{side}} A_{\text{side}}}{mc} \simeq \frac{9.804 \times 0.004\,07 + 9.615 \times 0.0139}{0.25 \times 4190} \simeq 0.000\,166.$$

Using this value in Equation (4.9) gives the predicted time for the tea to cool as $T \simeq \ln(62.5/42.5)/0.000\,166 \simeq 2330$ seconds.

The estimated time for the tea to cool to a drinkable temperature has changed from 2 hours 50 minutes to about 39 minutes. This is still too high, by a factor of three, but it is a significant improvement on the previous estimate.

To improve the estimate of T further, using this revised model, an even larger value for λ is needed. The greatest uncertainty in the value of λ is caused by the values of U and U_{side}, both of which depend on h_{tea} and h_{air}. In fact, these U-values are dominated by h_{air}, as you saw in one case in Exercise 4.9, and the greater the value of h_{air}, the greater the U-value. Suppose that h_{air}

takes its largest allowed value of 25. This gives $T \simeq 978$ seconds, that is, 16 minutes 18 seconds, which is still not very good as compared with the experimental time of 12 minutes 50 seconds. So again, in order to obtain a good estimate for T, the value for h_{air} would need to be outside its allowed range, which does not give much confidence in the model.

Once more this model gives us neither good quantitative results (the estimate for T) nor good qualitative results (the model still predicts exponential cooling), though the quantitative results are better than before.

Possible reasons for continuing discrepancies include the following.

- There may be draughts in the room, so that convection is forced rather than free, with the result that a value of h_{air} in the range 25–250 should be used.
- Radiative heat transfer may play a significant part in cooling the tea.
- Evaporation from the surface of the tea may play a significant part in cooling the tea.

See Table 4.2.

The modelling of evaporation is beyond the scope of this course, but you are asked to consider briefly the other two possibilities.

Exercise 4.13

Show that the value $h_{\text{air}} = 32\,\text{W}\,\text{m}^{-2}\,\text{K}^{-1}$ could be used with the revised model to obtain an estimate for the cooling time close to the value 12 minutes 50 seconds found by experiment. Would the revised model using this value for h_{air} give an adequate solution to the problem?

Exercise 4.14

According to a standard model, the rate of heat transfer by radiation is proportional to the difference between the fourth powers of the absolute temperatures (measured in kelvins, K) of the tea and of the air.

This model is described in MST209 *Unit 15*.

In broad terms, what would be the effect on the tea-cooling model of including radiative heat transfer? Would this help to satisfy the purposes of the model?

Exercise 4.13 shows that adjusting the value of h_{air} does not provide a satisfactory revision to the model. Exercise 4.14 suggests that the inclusion of radiative heat transfer will not improve the original model either. Instead, a different strategy is used in the next revision, to improve the model.

Second revision: two-compartment model

In this revision, the *first* model is taken as a starting point, and the concept of **compartmentalization** is introduced. Assumption (h) ignored the changes in the heat energy of the cup, but feeling the sides of a cup containing hot tea, it is apparent that some of the heat energy from the tea has transferred to the cup itself.

Treat the cup as a separate entity from the tea, each with its own temperature. As in the first model, conduction through the sides of the cup is ignored; instead, it is assumed that at any time t the temperature is uniform throughout the cup. However, both convection between the tea and the sides of the cup and convection between the sides of the cup and the surrounding air will be considered. There will be two stages to the transfer of heat energy: one for the heat energy transfer by convection from the tea to the air at the top and to the cup at the sides; and another for the heat energy transfer by convection from the sides of the cup to the surrounding air.

This is a *two-compartment model*, one compartment being the tea and the other the mug.

Exercise 4.15

After attempting each part of this exercise, you should check your answer to that part with the solution given at the back of the text before attempting the next part.

(a) How do the assumptions of the first model need to be modified, and what new assumptions need to be made?

(b) What additional variables and parameters are required for the revised model?

(c) Use the input–output principle, applied to the energy of the tea and of the cup over a small time interval $[t, t + \delta t]$, to determine a pair of first-order differential equations that describe the way in which the temperature of the cup and the temperature of the tea vary with time.

The assumptions of the first model are listed in the solution to Exercise 4.1, on page 55.

The variables and parameters for the first model are in Table 4.1 on page 23.

The mathematical model derived in Exercise 4.15 consists of a pair of *simultaneous linear first-order differential equations*, of the form

$$\begin{cases} \dot{\Theta} = -P\Theta + Q\Theta_{\text{cup}} + F, \\ \dot{\Theta}_{\text{cup}} = R\Theta - S\Theta_{\text{cup}} + G, \end{cases}$$

where P, Q, R, S, F and G are positive constants. You will see in Subsection 6.1 how the pair of differential equations obtained in Exercise 4.15 can be solved. The solution does go some way towards overcoming the qualitative deficiencies of the first model.

Simultaneous differential equations of this type are discussed and solved in MST209 *Unit 11*.

4.3 Commentary

This section introduced the modelling of heat transfer in a familiar setting. The first model developed gave an extremely poor quantitative prediction for the time taken for a cup of tea to cool to a drinkable temperature, more than 13 times the experimental outcome, and poor qualitative predictions on the way the tea cooled. Two revisions to this first model were considered, each based on a re-examination of the assumptions underpinning the first model.

The first revised model included heat losses by conduction and convection through the sides of the cup. While this gave some improvement in the quantitative prediction of the time taken for the tea to cool to a drinkable temperature (only three times as long as the experimental result), it did nothing to improve the qualitative prediction of the way the tea cooled. This was inevitable since this revised model, like the first model, assumed a linear rate of cooling and the data do not support this assumption.

The second revision was also based on the first model. Here conduction and radiation were ignored, but the heat energy of the cup was taken into account. Two objects (the tea and the cup) were considered independently to be possessing heat energy, and the energy transfer from one to the other and from each to the surrounding air was considered. The two-compartment model gives a reasonable fit to the data, as you will see in Subsection 6.1.

Even a fairly simple problem like this can generate a number of different avenues of investigation. The skill of the mathematical modeller is to try to tease out the essential mechanisms involved in any process, building on the success or failure of different attempts at modelling the process. Further revisions could be made to the models discussed here, for example, by combining the features of the two revisions that have been discussed. This

Two-compartment models (and many-compartment models) can be very useful in modelling a wide range of situations, including those where the physical attributes of an object are not uniformly distributed (as with the level of pollution in the Great Lakes, to be taken up once more in Section 5), and those where two or more separate objects are involved (as with the tea and cup here).

would lead to additional complexity, but may improve the model's ability to provide accurate predictions in a variety of different situations.

End-of-section Exercises

Exercise 4.16

Consider once more the assumption that the air remains at a constant temperature. For the experiment that was carried out to obtain data on the cooling of the tea, it was reported (alongside Table 4.3) that the air temperature was initially 18 °C, reducing to 17 °C by the end of the measurement period.

(a) If this variation is included in the first revised model, put forward a qualitative argument as to how it would affect the rate of cooling. How would the quantitative results change?

(b) Assuming a linear change in the air temperature from 18 °C to 17 °C in the time interval of 90 minutes, construct a mathematical relation for the air temperature as a function of time.

(c) Incorporate the relation you obtained in part (b) into the first revised model in Subsection 4.2.

5 A return visit to the Great Lakes

This section evaluates and revises the model, developed in Section 1, for pollution in the Great Lakes of North America. It is worth revisiting that section to remind yourself of the basic model and of the assumptions and simplifications that led to it. These are fundamental to a critical evaluation.

The information for this revised case study comes from two main sources:

The Great Lakes: An Environmental Atlas and Resource Book, published on the Internet at http://www.epa.gov/grtlakes/atlas/intro.html ;

Thomann, R.V. and Mueller, J.A. (1987) *Principles of Surface Water Quality Modeling and Control*, Harper and Row.

The first subsection below evaluates the success of the first model by comparing the model's predictions against data on contamination levels in the Great Lakes. The second subsection proposes some revisions to the first model in an attempt to overcome some of its deficiencies. The third subsection discusses briefly the techniques that were used, in order to highlight the key activities involved. In Subsection 6.1 you will have the opportunity to investigate some aspects of the revised models using the computer algebra package for the course.

Before going on to evaluate and revise the model, you should now view Part 2 of the video for this text. This features a discussion of some of the deficiencies of the model and provides suggestions for its revision.

Now view the second video sequence for this text, 'A second dip in the lakes'.

5.1 Evaluation of the model

To see if the model is realistic, consider its purpose. The problem is to predict how long it will take for the level of pollution in a lake to reduce to a target level if all sources of pollution are eliminated. It is intended to use the model to investigate pollution levels in any one of the Great Lakes, although the model could be used for any polluted lake.

The usual way of evaluating a model is to compare the predictions of the model with real data. These data can come either from your own experiments or from some published source. In the current case, collecting data from your own experiments is not feasible. However, there are useful published data, based on research studies conducted for the United States Environmental Protection Agency and the United States National Oceanographic and Atmospheric Administration by the University of Minnesota and the University of Wisconsin. One such set of data is shown in Table 5.1.

Table 5.1 Total PCB concentrations in Lake Superior

Year	PCB concentration (10^{-9} kg m^{-3})
1978	1.73
1979	4.04
1980	1.13
1983	0.80
1986	0.56
1988	0.33
1990	0.32
1992	0.18

PCB stands for polychlorinated biphenyl. PCBs have been used in the manufacture of transformers, capacitors and other electrical equipment.

The model considered pollution in general, but applies equally well to any single pollutant. The set of data in Table 5.1 is based on one particular type of pollutant, PCBs, which were a major source of pollution in the Great Lakes in the 1960s. The manufacture and importation of PCBs was prohibited in the United States in 1975. It is useful to consider this pollutant in the model since it is fairly stable, that is, it does not react with other chemicals, is not broken down by sunlight, and so on.

Exercise 5.1

Consider the data in Table 5.1.

(a) Why do you think that it is necessary to consider a stable pollutant?

(b) Is it relevant that the manufacture and importation of PCBs stopped in 1975?

(c) What are the advantages of using Lake Superior for the evaluation of the model?

A plot of these data reveals that there is a curious and unexpectedly high concentration in 1979. All of the other data points seem to lie close to a smooth curve. For the moment put this anomaly aside, and concentrate on the remaining data points. From the data it looks as if it takes 14 years (from 1978 to 1992, the actual time range of the data) for the concentration to fall to approximately a tenth of its former value, which relates to the specific question posed in the original model. However, the simple model predicts that it will take 439 years for the pollution level in Lake Superior to drop by a factor of ten. Clearly there is an enormous quantitative discrepancy between the model and these data.

Are the data at fault or is the model at fault? One might question the data, particularly on the basis of the unexpectedly high concentration given for 1979. Was this due to measurement error? Does this particular value indicate that the data may be unreliable? Does it really reflect the level of pollution in that year? Do PCBs actually satisfy the assumptions of the model or is there some process, relevant to PCBs, that has been overlooked? These are the kinds of questions that you should ask when using data to evaluate the success of a model.

Without going back over the data and checking how they were obtained, it is difficult to explain the anomalous value for 1979. In the video sequence it is suggested that this may have been caused by unusual disturbances of the sediment (for example, by storms, earthquakes or high winds). The sediment, which may contain high doses of the pollutant, would be drawn into the water and raise the measured value, while the reversion back to a much lower level the following year could have been caused by the PCBs settling back into the sediment from which they were disturbed. Regardless of whether this anomaly can be explained, these data will not validate the quantitative predictions of the model derived in Section 1.

The first model is not supported by the data. However, it is difficult to decide how to revise the model on the basis of a mismatch over a single value, namely, the time that it takes for the pollution concentration to reach a given level. A set of values or an equation to validate might be a better way to assess the success of the model and to give ideas on how best to revise it (as was the case when modelling the cooling of a cup of tea in Subsection 4.1).

The main equation in the simple model proposes a relationship between the pollutant concentration, at any time after all pollution has ceased, and the time:

$$c(t) = c(0)e^{-kt}, \quad \text{where } c(0) \text{ is the initial concentration of pollutant.}$$

Do the data bear any qualitative resemblance to this equation? Since it involves a negative exponential, taking the logarithm of both sides gives

$$\ln c(t) = \ln c(0) - kt,$$

which predicts a straight-line graph when $\ln c(t)$ is plotted against t.

The technique of taking logarithms in order to arrive at a predicted linear relationship was used in Subsection 4.1.

As Figure 5.1 illustrates, except for the anomalous value for 1979, the data are reasonably well approximated by a straight line. This confirms that the simple model does have some good qualitative agreement with the data, in the sense that a negative exponential model approximates reasonably well the data for PCBs. The value of k is given by the negative of the slope in Figure 5.1, that is, by $4.8 \times 10^{-9}\,\text{s}^{-1}$. The value of k derived from the model is $0.166 \times 10^{-9}\,\text{s}^{-1}$, which differs substantially from the experimental value, by a factor of approximately 30.

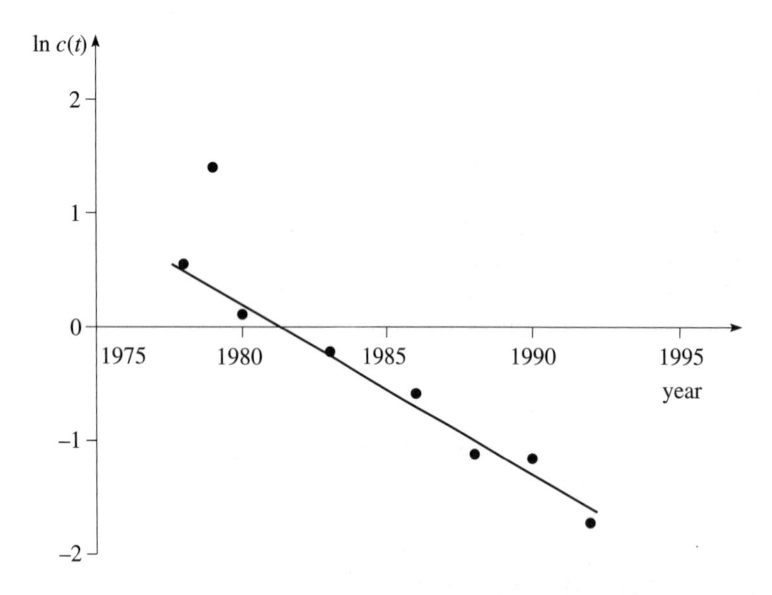

Figure 5.1 Log–linear plot of the data for Lake Superior

This is only one set of data and more sets of data ought to be tried. Unfortunately, there is little published data on the actual concentrations of pollutants in the lakes (their measurement is time-consuming and costly). However, there are extensive data available on contaminant concentrations in fish and birds inhabiting the Great Lakes. It has been shown (De Vault and Hesselberg, 1994) that, over the long term, trends of contaminants (including PCBs) in such inhabitants have followed those in the water and thus provide a measure of the trends in the Great Lakes' ecosystem. So it is reasonable to assume that if the simple model applies to the pollutant concentrations in the lakes, then it will apply to those in its inhabitants too.

De Vault, D. and Hesselberg, R. (1994) 'Contaminant trends in lake trout and walleye from the St Lawrence Great Lakes', *Journal of Great Lakes Research*.

Table 5.2 gives the concentrations of PCBs in herring gull eggs collected around Lake Superior over a number of years. The concentrations are given in micrograms per gram (or, equivalently, in parts per million), rather than in kilograms per cubic metre, since this is a more common method of measuring concentrations in solids.

To compare the concentration in herring gull eggs with the concentration in water, note that $1\,\mathrm{m}^3$ of water weighs 1000 kg. Hence in 1978 the concentration of PCBs in the water (from Table 5.1) was 1.73×10^{-12} grams of PCB per gram of water. In the herring gull eggs that year the concentration was 44×10^{-6} grams of PCB per gram of egg, which is a magnification of about 25 million.

Table 5.2 PCB concentrations in herring gull eggs from Lake Superior

Year	PCB concentration ($\mu\mathrm{g\,g}^{-1}$)	Year	PCB concentration ($\mu\mathrm{g\,g}^{-1}$)
1974	65	1985	21
1975	81	1986	17
1977	59	1987	14
1978	44	1988	15
1979	60	1989	17
1980	28	1990	13
1981	36	1991	15
1982	37	1992	15
1983	23	1993	16
1984	19		

The high concentrations here as compared with those in the lake are an example of *biomagnification*, in which small concentrations of pollutant at one point cause higher and higher concentrations when moving up the food chain.

Exercise 5.2

(a) When evaluating the model, why should you disregard at least the first data point in Table 5.2?

(b) Why do you think that the concentrations are rising towards the end?

Based on Exercise 5.2, it seems reasonable to eliminate the first data point and the last three data points from Table 5.2. This gives a value for k of 3.8×10^{-9}, which is similar to the value obtained with the first set of data. This confirms that, while the negative exponential form of the relationship is approximately correct, the actual numerical values predicted by the model are in error.

The comparison being made here is valid provided that the magnification factor (from concentration in the water to concentration in herring gull eggs) and the time delay (referred to in the solution to Exercise 5.2(a)) are both constant.

Whereas the quantitative comparison (between the data and the model) of the time for the concentration of pollutant to drop to a tenth of its initial value gives no clue how to revise the model, the qualitative comparison gives some information on where to focus revision of the model: in the assumptions concerning the rate at which the pollutant dissipates.

5.2 Revisions to the model

This subsection investigates two possible revisions to the model as applied to PCBs. The first revision takes account of a process that affects the rate at which PCB concentration diminishes. The second revision tries to take into account the variation in PCB concentration across the lake.

In considering revisions to the model, first review the modelling assumptions.

Exercise 5.3

Consider the assumptions made for the first model. Which of these assumptions do you think should be re-examined?

These assumptions are listed on page 7.

First revision: adding a process

To explain the discrepancy between the prediction of the model and evidence of the data, the physics and chemistry of PCBs need examination, to determine the likeliest ways in which PCB concentrations can diminish other than by being flushed out of the lake by the flow of water. For any chemical in solution, there are four principal mechanisms which can cause its concentration to diminish:

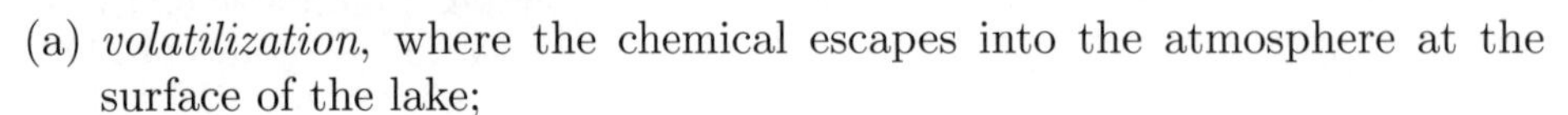

(a) *volatilization*, where the chemical escapes into the atmosphere at the surface of the lake;

(b) *photolysis*, where the action of sunlight on water near the surface of the lake causes certain chemicals to degrade;

(c) *hydrolysis*, where the chemical reacts with water, causing it to break down;

(d) *biodegradation*, where the chemical is degraded by bacteria and other living organisms.

Now PCBs are stable. This means that they are difficult to break down, which tends to rule out photolysis, hydrolysis and biodegradation as relevant processes. However, there is evidence that PCBs are volatile, and so this mechanism is considered further. Inclusion of volatilization amends Assumption (b), as follows:

(b) the pollutant does not biodegrade in the lake or decay through any other biological, chemical or physical process *except through volatilization.*

An additional assumption is also needed, about the process of volatilization.

Exercise 5.4

Upon what physical properties does the rate of volatilization depend?

For a simple revision of the model, consider only the first two factors mentioned in the solution to Exercise 5.4. This results in the following additional assumption:

(h) *the pollutant is lost from the lake to the atmosphere at a rate that is proportional to the total surface area of the lake and to the concentration of pollutant in the lake.*

Suppose that the surface area of the lake is A (measured in square metres). The assumption is that the mass of the pollutant is lost at a rate that is proportional both to A and to $c(t)$, where $c(t)$ is the pollutant concentration in the lake. Hence pollutant is lost at a rate proportional to $Ac(t)$. Thus the differential equation that models the rate of loss of the mass of pollutant is

$$\frac{dm}{dt} = -km(t) - pAc(t), \tag{5.1}$$

This is a generalization of Equation (1.1). A detailed derivation of this differential equation would involve use of the input–output principle.

where p is the constant of proportionality, known as the *volatilization rate.* The mass–concentration relationship given previously, $c(t) = m(t)/V$, is still valid. So, by eliminating $c(t)$, Equation (5.1) can be expressed entirely in terms of the mass m of pollutant, as

$$\frac{dm}{dt} = -km(t) - \frac{pA}{V}m(t) = -\left(k + \frac{pA}{V}\right)m(t) = -\kappa m(t), \tag{5.2}$$

where $\kappa = k + pA/V$. This is essentially the same model as the first one, except that the proportionate flow rate k has now become a *proportionate decay rate* κ, and κ should be greater than k.

The solution to Equation (5.2) is

$$m(t) = m(0)e^{-\kappa t}, \quad \text{where } m(0) \text{ is the initial mass of pollutant.}$$

The corresponding equation for the pollutant concentration is

$$c(t) = c(0)e^{-\kappa t}, \quad \text{where } c(0) \text{ is the initial concentration of pollutant.}$$

The time, T, taken for the pollutant concentration to reduce to the target level $c(T) = c_{\text{target}}$ is given by

$$T = -\frac{1}{\kappa} \ln\left(\frac{c_{\text{target}}}{c(0)}\right).$$

The proportionate decay rate $\kappa = k + pA/V$ determines the value of T.

Now this model is used to predict what will happen to PCB concentrations in Lake Superior. It is a negative exponential decay model, and therefore will have the same desirable qualitative behaviour that was apparent in the first model. Since the proportionate decay rate κ is larger than the proportionate flow rate k, it is also likely to give better quantitative agreement than the first model.

Thomann and Mueller (1987) suggest that volatilization of PCBs probably occurs at a rate of about 0.1 m (that is, $0.1\,\text{m}^3$ per m^2) per day. So, in SI units, $p \simeq 0.1/(24 \times 60 \times 60) \simeq 1.16 \times 10^{-6}\,\text{m}\,\text{s}^{-1}$. The surface area of Lake Superior is given as $A \simeq 8.21 \times 10^{10}\,\text{m}^2$. Hence, in SI units,

Thomann, R.V. and Mueller, J.A. (1987) *Principles of Surface Water Quality Control Modeling*, Harper and Row.

$$\begin{aligned}\kappa &= k + pA/V \\ &\simeq 0.166 \times 10^{-9} + 1.16 \times 10^{-6} \times 8.21 \times 10^{10}/(12.10 \times 10^{12}) \\ &\simeq 8.02 \times 10^{-9}.\end{aligned}$$

The estimated proportionate decay rate is about 50 times larger than the proportionate flow rate. It is therefore clear that the volatilization of PCBs has a much greater influence on the rate at which their concentration is reduced than does the flow of pollutant out of the lake. The revised time, in seconds, for the pollutant concentration to fall by a factor of ten is

$$T = \frac{\ln 10}{\kappa} \simeq 2.87 \times 10^8,$$

which is approximately 9.1 years.

The results of this revised model are much more encouraging. The values in Table 5.1 indicate that PCB concentrations should take about 14 years to reduce by a factor of ten, while the revised model predicts about 9 years (compared with the 439 years predicted by the original model). So, from a model that grossly overestimates the time taken, the revised model underestimates it. It may be that the revised model is adequate. If it is not, then possible reasons for the discrepancy between the revised model's prediction and reality need to be considered.

Possible reasons for this discrepancy include the following.

- The volatilization rate is only given to one significant figure, and this may cause a sizeable numerical error. A more accurate estimate for this rate is required to ensure reliable predictions.
- The volatilization of PCBs from the lake to the atmosphere has been included, but the absorption of PCBs from the atmosphere to the lake has been ignored.

- Although PCBs were banned in 1975, there will be a residual amount in the catchment area that will continue to wash into the lake over time, so Assumption (a) may need to be revised.
- Some of the PCBs settle into the sediment at the bottom of the lake, and will be flushed out of this sediment at a much slower rate.
- The volatilization rate is based on the assumption that the pollutant is uniformly distributed over the whole lake. However, it is known that the pollutant concentration varies considerably within the lake, and so this assumption may have to be reconsidered.

The first item in this list is not a criticism of the model but a requirement for more accurate data. Each of the other criticisms could form the basis of a second revised model.

The process of improving a model is one of evaluating the model at each stage and then addressing the assumption that seems most likely to improve the model; it is an iterative process. It is best to change one feature at a time, even though that may affect more than one assumption.

In the second revision of the model below, the assumption of a uniform concentration of pollutant in the whole lake is revised, to take some account of the last criticism.

Second revision: segmenting the lake

To account for the variation in pollutant concentration across a lake, break down the region considered by the model into a number of smaller subregions or cells, with each cell having its own distinct properties. A model for the whole region can then be built up by considering what happens in each cell, and how it interacts with the other cells.

This technique, known as *segmentation* or *compartmentalization*, was used in a different way in Subsection 4.2.

You saw, on the video sequence, that the mathematical model for pollution in the Great Lakes used by the US Environmental Protection Agency considers what happens on each lake by segmenting it into a large number of smaller volumes. This segmentation is done by first constructing a horizontal grid on the lake. The horizontal segmentation takes into account, for example, regions of the lake where the pollutant concentration is much higher or lower than average. Then, for each cell created by the horizontal segmentation, there is also vertical segmentation. This allows for differences in behaviour between the upper regions of the lake, where photolysis and volatilization take place, and the lower regions, where the presence of sediment may have an effect. In this way each lake is divided up into hundreds of small cells, on which more precise data are available and for which more precise information may be required. Each cell has its own variables representing, for example, its pollutant concentration and the rates of flow to and from adjacent cells.

Models that use segmentation can be very complicated, and this is true of the models used to analyse levels of pollution in the Great Lakes. So, to keep the analysis relatively simple, the revised model here is a simple segmentation of Lake Huron. In the south-west of Lake Huron is a large bay, called Saginaw Bay (see Figure 5.2). The Saginaw River flows into this bay and was a major source of pollution for the whole lake. Suppose that a model of the concentration of PCBs in Lake Huron is required. It would be reasonable to consider Saginaw Bay separately from the rest of the lake, as shown in Figure 5.2, since the concentration of pollutant in this bay is likely to be significantly higher than that in the rest of the lake. Two further

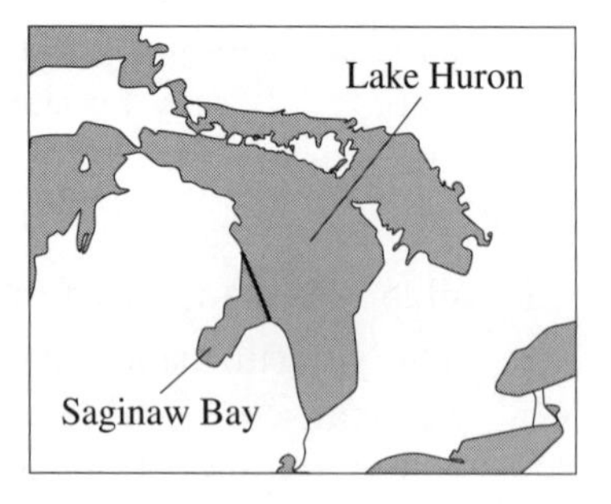

Figure 5.2

assumptions are now needed, in addition to those made for the first revised model:

(i) *Saginaw Bay and the rest of Lake Huron can be considered as two separate lakes;*

(j) *the rate of diffusion of pollutant from Saginaw Bay into the rest of Lake Huron is proportional to the difference in concentrations between the two regions of the lake.*

Hence this is a two-compartment model. You saw a similar model in Subsection 4.2.

Diffusion of pollutant is a transportation process that involves the movement of molecules of the pollutant relative to the water, rather than bodily movement of the water itself.

The first assumption here implies that the pollutant in Saginaw Bay is evenly dispersed throughout the bay, so that the concentration of pollutant in the bay varies only with time and not with position. The second assumption models the interaction between the two regions of the lake.

There will be approximately twice as many variables and parameters as before, with one set for each region of the lake.

The variables are:

t the time, in seconds, since all sources of pollution ceased;
$m_{\mathrm{B}}(t)$ the mass, in kilograms, of pollutant in the bay at time t;
$m_{\mathrm{L}}(t)$ the mass, in kilograms, of pollutant in the rest of the lake at time t;
$c_{\mathrm{B}}(t)$ the concentration, in kilograms per cubic metre, of pollutant in the bay at time t;
$c_{\mathrm{L}}(t)$ the concentration, in kilograms per cubic metre, of pollutant in the rest of the lake at time t.

The parameters are:

V_{B} the volume, in cubic metres, of water in the bay;
V_{L} the volume, in cubic metres, of water in the rest of the lake;
r_{B} the water flow rate, in cubic metres per second, into and out of the bay;
r_{L} the water flow rate, in cubic metres per second, into and out of the rest of the lake;
A_{B} the surface area, in square metres, of the bay;
A_{L} the surface area, in square metres, of the rest of the lake;
p the volatilization rate, in metres per second, for the pollutant;
E the bulk exchange rate, in cubic metres per second.

The meaning of E will become apparent shortly.

Starting with Saginaw Bay, consider what happens to the mass of pollutant in the bay over a time interval $[t, t+\delta t]$. According to Assumptions (a) and (g) there is no new pollution entering the lake, so there is no input of pollutant to the bay. The output (in kilograms per second) of pollutant from the bay takes three main forms.

- Pollutant flows from Saginaw Bay into the rest of Lake Huron, at a rate of $r_{\mathrm{B}}c_{\mathrm{B}}(t) = r_{\mathrm{B}}m_{\mathrm{B}}(t)/V_{\mathrm{B}}$.
- Pollutant is lost to the atmosphere, through volatilization, at a rate of $pA_{\mathrm{B}}c_{\mathrm{B}}(t) = pA_{\mathrm{B}}m_{\mathrm{B}}(t)/V_{\mathrm{B}}$.
- Pollutant diffuses from a region of higher concentration (the bay) to a region of lower concentration (the rest of the lake) at a rate of $E(c_{\mathrm{B}}(t) - c_{\mathrm{L}}(t)) = E(m_{\mathrm{B}}(t)/V_{\mathrm{B}} - m_{\mathrm{L}}(t)/V_{\mathrm{L}})$, where the parameter E depends, for example, on the vertical cross-sectional area separating Saginaw Bay from the rest of Lake Huron.

This follows from Assumption (j). The parameter E is known as the *bulk exchange rate* for diffusion of pollutant across the boundary between the bay and the rest of the lake.

Thus the total output of pollutant from the bay during the time interval $[t, t+\delta t]$ is

$$\left(\frac{r_{\mathrm{B}}}{V_{\mathrm{B}}}m_{\mathrm{B}}(t) + \frac{pA_{\mathrm{B}}}{V_{\mathrm{B}}}m_{\mathrm{B}}(t) + \frac{E}{V_{\mathrm{B}}}m_{\mathrm{B}}(t) - \frac{E}{V_{\mathrm{L}}}m_{\mathrm{L}}(t)\right)\delta t.$$

The accumulation of the mass of pollutant within the bay is

$$m_B(t + \delta t) - m_B(t).$$

Using the input–output principle then gives

$$m_B(t+\delta t) - m_B(t) = -\left(\frac{r_B}{V_B}m_B(t) + \frac{pA_B}{V_B}m_B(t) + \frac{E}{V_B}m_B(t) - \frac{E}{V_L}m_L(t)\right)\delta t.$$

Dividing through by δt, and then letting $\delta t \to 0$, leads to the differential equation

$$\frac{dm_B}{dt} = -k_1 m_B(t) + k_2 m_L(t), \qquad (5.3)$$

where $k_1 = r_B/V_B + pA_B/V_B + E/V_B$ and $k_2 = E/V_L$.

Consider now the rest of Lake Huron. The input (in kilograms per second) of pollutant to this part of the lake comes from two main sources:

- the flow of polluted water from Saginaw Bay, at a rate of $r_B m_B(t)/V_B$;
- diffusion across the boundary between the two parts of the lake, at a rate of $E(m_B(t)/V_B - m_L(t)/V_L)$.

This ignores any input of pollutant to Lake Huron from the upstream lakes, in keeping with Assumption (g). This assumption may not be justified for Lake Huron, but it keeps the model simple.

Hence the input of pollutant to the rest of Lake Huron in the time interval $[t, t + \delta t]$ is

$$\left(\frac{r_B}{V_B}m_B(t) + \frac{E}{V_B}m_B(t) - \frac{E}{V_L}m_L(t)\right)\delta t.$$

During this same time interval, the output of pollutant from this part of the lake takes two main forms.

- Pollutant flows from Lake Huron to Lake Erie, at a rate of $r_L c_L(t) = r_L m_L(t)/V_L$.
- Pollutant is lost to the atmosphere, through volatilization, at a rate of $pA_L c_L(t) = pA_L m_L(t)/V_L$.

Thus the total output of pollutant from the rest of the lake in the time interval $[t, t + \delta t]$ is

$$\left(\frac{r_L}{V_L}m_L(t) + \frac{pA_L}{V_L}m_L(t)\right)\delta t.$$

The accumulation of the mass of pollutant within the rest of the lake is

$$m_L(t + \delta t) - m_L(t).$$

Using the input–output principle, applied to the rest of Lake Huron, then leads to the differential equation

$$\begin{aligned}\frac{dm_L}{dt} &= \frac{r_B}{V_B}m_B(t) + \frac{E}{V_B}m_B(t) - \frac{E}{V_L}m_L(t) - \frac{r_L}{V_L}m_L(t) - \frac{pA_L}{V_L}m_L(t)\\ &= k_3 m_B(t) - k_4 m_L(t), \qquad (5.4)\end{aligned}$$

where $k_3 = r_B/V_B + E/V_B$ and $k_4 = E/V_L + r_L/V_L + pA_L/V_L$.

The revised model has led to Equations (5.3) and (5.4). These are a pair of linear differential equations that describe the behaviour of pollutant in the two regions of the lake. The solution of these equations, which forms part of 'Do the mathematics' for this model, will be addressed in Subsection 6.1.

Techniques for solving systems of linear differential equations are to be found in MST209 *Unit 11*.

Exercise 5.5

Before tackling the solution of this second revised model, it is worthwhile estimating the contribution from each term that arose in applying the input–output principle. Table 5.3 gives relevant data on parameter values for the two regions of the lake.

(a) Suppose that the concentration of pollutant in the rest of the lake is half that in Saginaw Bay. Compare the rates of pollutant output from the bay for each of the three forms of pollutant transfer: water flow, volatilization and diffusion. Hence decide which of these mechanisms is the most significant in removing pollutant from the bay.

(b) How will the rate of pollutant output for each of these three forms change, if the concentration of pollutant in the rest of the lake is less than half that in the bay?

(c) What is the main mechanism for the removal of pollutant from the rest of the lake?

Table 5.3

V_B	$25 \times 10^9\,m^3$
A_B	$4.2 \times 10^9\,m^2$
r_B	$153\,m^3\,s^{-1}$
V_L	$3.24 \times 10^{12}\,m^3$
A_L	$57 \times 10^9\,m^2$
r_L	$4967\,m^3\,s^{-1}$
p	$1.16 \times 10^{-6}\,m\,s^{-1}$
E	$11\,000\,m^3\,s^{-1}$

5.3 Commentary

In this section you have seen how data were used to evaluate a simple model for the behaviour of pollution in the Great Lakes. The data were obtained from published sources, but it was not possible to check their reliability. There was good qualitative agreement with the model but rather poor quantitative agreement.

Prior to making revisions, the assumptions made for the first model were evaluated. The first revision was based on the inclusion of an additional process, to try to explain the reason for the substantial quantitative differences between the published data and the predictions of the model. This revision was very effective, since it demonstrated that volatilization is likely to have a much more substantial effect on the removal of PCBs than does the flow of water out of the lake.

The second revision was based on a segmentation of the lake, or division into compartments, to take into account the non-uniformity of the concentration of pollutant. This technique is widely applicable in mathematical modelling. Exercise 5.5 shows that diffusion, as well as volatilization, is more significant than water flow in removing pollutant from Saginaw Bay. In fact, the only form of pollutant output considered in the original model turns out to be the least important in the revised models.

The original model, based on reasonable simplifying assumptions, made a start to the problem-solving process. The evaluation of this model revealed its deficiencies, and as a result other forms of output were considered to improve the model. The development that was undertaken here emphasizes the role of the first simple model in kick-starting the modelling process.

6 Computer explorations

Subsection 6.1 is a session applying the computer algebra package, while Subsection 6.2 is based on the multimedia package for the text.

6.1 Solving systems on the computer

In this subsection you will see how the solutions of models developed earlier in the text can be explored using the computer algebra package.

Each activity has its own associated worksheet. An introduction and activity statement are given in the worksheet. In some cases the symbols for the variables on the worksheet may differ slightly from the ones used here, but each symbol is defined in the worksheet. After carrying out each part of an activity, read the corresponding 'Comment' in the worksheet. Then read the 'Final remarks', if any.

The activity statements in this text are expanded versions of those to be found in the worksheets (usually to include material from the worksheet introduction).

Use your computer to carry out the following activities.

Activity 6.1

Sensitivity of skid marks model

This activity looks at the sensitivity of the skid marks model, developed in Subsection 2.1, to changes in data. To estimate the initial speed of a car that has skidded to a halt, a test car with similar tyres performs a controlled skid. The model from Subsection 2.1 is extended here so that two examples of how the solution might be sensitive to changes in the data can be considered.

- The model from Subsection 2.1 assumes that the deceleration of the original car is the same as the deceleration of the test car. However, the deceleration depends on the state of the tyres, the wetness of the road surface, and so on. If it is not possible to perform the test skid under the same conditions as the original skid, then the assumption of equal decelerations is not valid. Therefore, the extended model here includes a parameter α so that the deceleration of the original car relative to the test car can be varied, by putting $a_{\text{car}} = \alpha a_{\text{test}}$.
- The first model assumes that the original car comes to a complete stop under braking. If there is a collision then this assumption may not be valid. The extended model includes the final speed of the car, so that the effect of stopping due to collision can also be examined.

The speed of the car at the start of the skid will thus be estimated as

$$u_{\text{car}} = \sqrt{v_{\text{car}}^2 + \alpha u_{\text{test}}^2 \frac{x_{\text{car}}}{x_{\text{test}}}},$$

This equation follows from the solution to Exercise 2.1(o), with

$$a_{\text{test}} = -\frac{u_{\text{test}}^2}{2x_{\text{test}}}$$

as before.

where v_{car} is the final speed of the car, u_{test} is the speed of the test car at the start of its skid, and x_{car} and x_{test} are the two skid lengths, measured using the same units.

The example considered has $x_{\text{car}} = 55.2$ yards, $x_{\text{test}} = 29.2$ yards and $u_{\text{test}} = 50$ mph. The road has a speed limit of 70 mph.

(a) Assume that $\alpha = 1$ and that the final speed of the original car is zero. Determine whether the car has broken the speed limit.

(b) The road has begun to dry out, and the tyres of the test car are significantly better than the tyres on the crashed car. It is estimated that the appropriate value of α lies between 0.6 and 0.8. Estimate the range of initial speeds that could have led to these skid marks.

(c) Rain has begun to fall in the time since the car crashed. It is estimated that the appropriate value of α lies between 1.5 and 2. Estimate the range of initial speeds that could have led to these skid marks.

(d) Consider the case where the original car crashed into the back of a stationary car. It is estimated that the final speed at the moment of impact was between 5 and 10 mph. For $\alpha = 1$, estimate the range of initial speeds that could have led to these skid marks.

(e) You are employed by the car driver's solicitor. The driver has been charged with speeding. Given that α has been estimated to lie between 1.5 and 2 and that, at the moment of impact, the final speed of the car was between 5 and 10 mph, estimate the range of initial speeds that could have led to these skid marks. Comment on your findings.

Activity 6.2

Two-compartment model for tea cooling

In Subsection 4.2, the second revised model for the cooling of a cup of tea was derived as

$$\begin{cases} \dot{\Theta} = -P\Theta + Q\Theta_{\text{cup}} + F, \\ \dot{\Theta}_{\text{cup}} = R\Theta - S\Theta_{\text{cup}} + G, \end{cases}$$

See Equations (S.8) and (S.9) in the solution to Exercise 4.15(c).

where $\Theta(t)$ and $\Theta_{\text{cup}}(t)$ are respectively the temperature of the tea and the cup at time t, and

$$P = \frac{UA + h_{\text{tea}}A_{\text{side}}}{mc}, \quad Q = \frac{h_{\text{tea}}A_{\text{side}}}{mc}, \quad F = \frac{UA}{mc}\Theta_{\text{air}},$$

$$R = \frac{h_{\text{tea}}A_{\text{side}}}{m_{\text{cup}}c_{\text{cup}}}, \quad S = \frac{(h_{\text{tea}} + h_{\text{air}})A_{\text{side}}}{m_{\text{cup}}c_{\text{cup}}}, \quad G = \frac{h_{\text{air}}A_{\text{side}}}{m_{\text{cup}}c_{\text{cup}}}\Theta_{\text{air}}.$$

The parameters here are as listed in Table 4.1 on page 23 and in the solution to Exercise 4.15(b). Note, however, that the worksheet for this activity makes the following changes of notation:

The variables and parameters used in the worksheet are listed in its introduction.

$$\Theta_{\text{tea}} \text{ for } \Theta, \quad m_{\text{tea}} \text{ for } m, \quad c_{\text{tea}} \text{ for } c, \quad A_{\text{top}} \text{ for } A.$$

The subscripts 'tea' and 'top' here could have been introduced in Subsection 4.2. However, the symbols Θ, m, c and A stood there for the same quantities as in the original model, in Subsection 4.1, where subscripts would have been superfluous.

Also Θ_0 (the initial temperature of the tea) becomes $\Theta0_{\text{tea}}$, and the initial temperature of the cup is denoted by $\Theta0_{\text{cup}}$. Other symbols used are t_{exp}, for the time in minutes $(=\frac{1}{60}t)$, and Θ_{exp} for the corresponding experimental temperature values (as in Table 4.4).

The following data are used for this model:

$$m_{\text{tea}} = 0.25, \quad c_{\text{tea}} = 4190, \quad m_{\text{cup}} = 0.25, \quad c_{\text{cup}} = 920,$$

$$h_{\text{air}} = 10, \quad h_{\text{tea}} = 500, \quad U = (1/h_{\text{air}} + 1/h_{\text{tea}})^{-1}, \quad \Theta_{\text{air}} = 17.5,$$

$$D = 0.072, \quad H = 0.0615, \quad A_{\text{top}} = \tfrac{1}{4}\pi D^2, \quad A_{\text{side}} = \pi DH.$$

(a) Solve the system of equations given that, at time $t = 0$, $\Theta_{\text{tea}} = 80$ and $\Theta_{\text{cup}} = \Theta_{\text{air}}$. Determine the time that the tea takes to cool to 60 °C.

(b) Compare your answers, both quantitatively and qualitatively, with the data given in Table 4.4.

(c) Vary the values of h_{tea} and h_{air} in the ranges $50 \le h_{\text{tea}} \le 1000$ and $2 \le h_{\text{air}} \le 25$, to see if you can improve the qualitative and quantitative match with the data.

Activity 6.3

Fitting data for Great Lakes models

Both the original model for pollution in the Great Lakes (Section 1 and Subsection 5.1) and the first revision to it (Subsection 5.2) led to a prediction that the pollutant concentration $c(t)$ and time t are related by an equation of the form

$$\ln c(t) = \ln c(0) + \lambda t, \qquad (6.1)$$

where λ is a constant (the slope of the straight line obtained by plotting $\ln c(t)$ against t). In fact, $-\lambda$ is equal to the proportionate flow rate k in the original model, and to the proportionate decay rate κ in the first revised model. The connection between k and κ is

$$\kappa = k + \frac{pA}{V},$$

where A and V are respectively the area and volume of the lake, and p is the volatilization rate, taken to be $1.16 \times 10^{-6}\,\mathrm{m\,s^{-1}}$.

On page 39 you saw how values of A, V and k for Lake Superior led to values for κ and for the decay time T (the time for pollutant concentration to fall by a factor of ten). Table 6.1 below gives the corresponding results for κ and decay time for all of the lakes.

Table 6.1 Proportionate flow and decay rates, and times to fall by factor of ten

			Original model		Revised model	
Lake	Volume V (10^{12} m^3)	Surface area A (10^{10} m^2)	Proportionate flow rate k (10^{-9} s^{-1})	Decay time (years)	Proportionate decay rate κ (10^{-9} s^{-1})	Decay time (years)
Superior	12.10	8.21	0.166	439	8.02	9.1
Michigan	4.92	5.78	0.319	229	13.92	5.2
Huron	3.54	5.96	1.441	51	20.93	3.5
Erie	0.48	2.57	12.292	6	74.26	1.0
Ontario	1.64	1.90	4.134	18	17.54	4.2

In this activity, you are asked to investigate whether data on pollution levels in the Great Lakes are in qualitative agreement with the straight-line relationship predicted by Equation (6.1) and, if so, whether the value of $-\lambda$ obtained matches the value of k or κ in Table 6.1. You are also asked to assess whether the data match the predicted times for pollutant concentrations to reduce by a factor of ten, as given in Table 6.1.

The values of k in Table 6.1 were obtained, from the water flow rate and volume of each lake, in Table 1.1. The decay time is given (in seconds) by

$$T = (\ln 10)/k$$

in the original model, and by

$$T = (\ln 10)/\kappa$$

in the revised model.

The worksheet contains data on:

- total PCB concentrations for Lake Superior;
- PCB concentrations in lake trout for Lake Michigan;
- PCB concentrations in herring gull eggs for all five of the Great Lakes.

Some of these data are from before 1975, when the introduction of PCBs was banned from the Great Lakes.

For each of the seven sets of data, do the following.

(a) Remove any anomalous data values at either end of the data set.

(b) For the remaining data, plot $\ln c(t)$ against t and a best-fit straight line.

(c) Determine the slope λ of this straight line, and compare the value of $-\lambda$ with the appropriate values of k and κ in Table 6.1.

(d) Determine the time T, predicted by the straight-line plot, for the pollutant concentration to reduce by a factor of ten, and compare it with the values for decay time in Table 6.1.

Hence, for each lake, decide whether either model satisfies the purpose of predicting the variation in pollution levels in the lake and, in particular, of predicting the time taken for the concentration of pollutant to reduce by a factor of ten.

Activity 6.4

Two-compartment model for Lake Huron

In Subsection 5.2, the second revised model for PCB levels in Lake Huron is based on segmenting the lake into two regions, with Saginaw Bay treated separately from the rest of the lake. The model for the masses of pollutant at time t, in Saginaw Bay ($m_B(t)$) and in the rest of the lake ($m_L(t)$), is given by the equations

$$\frac{dm_B}{dt} = -k_1 m_B(t) + k_2 m_L(t), \tag{5.3}$$

$$\frac{dm_L}{dt} = k_3 m_B(t) - k_4 m_L(t), \tag{5.4}$$

where

$$k_1 = \frac{r_B + pA_B + E}{V_B}, \quad k_2 = \frac{E}{V_L}, \quad k_3 = \frac{r_B + E}{V_B}, \quad k_4 = \frac{E + r_L + pA_L}{V_L}.$$

The parameters of the model are defined using SI units, and their values are given below.

The surface area, volume and water flow rate for Saginaw Bay:

The definitions of these parameters are on page 41. Their values are as in Table 5.3.

$$A_B = 4.2 \times 10^9, \quad V_B = 25 \times 10^9, \quad r_B = 153.$$

The surface area, volume and water flow rate for the rest of Lake Huron:

$$A_L = 57 \times 10^9, \quad V_L = 3240 \times 10^9, \quad r_L = 4967.$$

The bulk exchange rate between the two parts of Lake Huron: $E = 11\,000$.

The volatilization rate for the pollutant: $p = \dfrac{0.1}{24 \times 60 \times 60} = 1.16 \times 10^{-6}$.

Assume that this model has been validated, and gives a reasonable model for PCB contamination.

Suppose that, initially, there is no PCB pollution in Lake Huron and that, at time $t = 0$, 1000 kg of PCBs are dumped into Saginaw Bay. The recommended maximum concentration of PCBs in the water, for the protection of wildlife, is $17 \times 10^{-12}\,\mathrm{kg\,m^{-3}}$.

(a) How long will it be before this maximum concentration of PCBs is reached in the rest of Lake Huron?

(b) How long will it take for the rest of Lake Huron to reach its peak PCB concentration, and what is this level of PCB concentration?

(c) How long will it be before the water in the rest of Lake Huron returns to a safe level of contamination?

(d) How long will it be before Saginaw Bay returns to a safe level of contamination?

6.2 Back to the Great Lakes once more

This subsection revisits once again the topic of pollution levels in the Great Lakes and uses a multimedia package to look at three different models in this context. Each model features a system of linear constant-coefficient first-order differential equations.

- The first model treats the five lakes as a single system and analyses the passage of a pollutant through this five-compartment system.
- The second model reverts to a single lake, but includes consideration of the pollutant transfers that occur between the water and the underlying sediment of the lake.
- The third model returns to a situation considered in Subsection 5.2 and in Activity 6.4, where pollutant is exchanged between Saginaw Bay and the main body of Lake Huron, but adds the extra feature of an input to the bay from the Saginaw River. This leads to an inhomogeneous system of differential equations, whereas the systems for the two earlier models are homogeneous.

The package also explains how the eigenvalues and eigenvectors of the coefficient matrix can be used to solve many homogeneous systems of linear constant-coefficient first-order differential equations. An interactive animation is included which gives insight into the geometric meaning of eigenvalues and eigenvectors.

This is an alternative explanation to that provided in MST209 *Unit 11*.

Now work through the second multimedia package for this text, 'Revising the model for the Great Lakes'.

Outcomes

After studying this text you should be able to:

- create simple models, given a clear statement of the problem;
- write down the simplifying assumptions that underpin a model;
- identify the key variables and the parameters of a model;
- apply the input–output principle to obtain a mathematical model, where appropriate;
- obtain mathematical relationships between variables, based on or linking back to the simplifying assumptions;
- interpret the mathematical solution to a modelling problem in terms of the original statement of the problem;
- understand the processes involved in evaluating a model, both qualitatively and quantitatively;
- appreciate the role of data in testing the model and, if necessary, in providing parameter values for the model;
- understand that the purpose of a model is the benchmark used to judge the suitability of the model;
- appreciate that simple models can be as useful as more complex models, if they serve their intended purpose;
- appreciate the role of the assumptions of the model in thinking about possible revisions;
- appreciate that there are a number of possible revisions to a model;
- use a variety of techniques to revise a simple model;
- appreciate that a complex model may be built up from a simple model by gradually including more and more features;
- investigate the sensitivity of a solution to small changes in the data, using a computer package.

Solutions to the exercises

Section 1

1.1 In describing the formulation of the model, it may be a help to write down some words or phrases first, and then to combine them into a paragraph.

Possible key words or phrases are: pollution has ceased; the water flowing from the lake is polluted; concentration of the pollutant is uniform; small time interval; mass of pollutant; one lake; no change in volume of the lake; input–output principle.

A possible description might read:

> This model considers a polluted lake to which no further pollutant is added. The lake is of constant volume. The water flowing from the lake is polluted, and this is the only way that the pollutant leaves the lake. The input–output principle is applied to the mass of pollutant within the lake, based on the change over a small time interval.

In a modelling report, the description of the formulation should be placed before the actual formulation, in order to guide the reader. However, it should actually be written last, after the formulation has been completed.

1.2 **(a)** The input–output principle is applied to the mass of pollutant within the lake. In the time interval $[t, t+\delta t]$, the input (mass entering the lake) is $q\,\delta t$, while the output is $(r/V)m(t)\delta t$, as shown in the previous text. The accumulation over this time interval is $m(t+\delta t) - m(t)$. Hence the change in the mass of pollutant in this interval is given by

$$m(t+\delta t) - m(t) \simeq q\,\delta t - \frac{r}{V}m(t)\delta t.$$

Putting $k = r/V$, this leads to the differential equation

$$\frac{dm}{dt} = q - km(t). \qquad \text{(S.1)}$$

(b) The integrating factor method gives the general solution of Equation (S.1) as

$$m(t) = \frac{q}{k} + Ce^{-kt},$$

where C is an arbitrary constant. Rearranging the solution, and using the initial condition at $t = 0$ to evaluate C, gives the required result,

$$m(t) = \left(m(0) - \frac{q}{k}\right)e^{-kt} + \frac{q}{k}. \qquad \text{(S.2)}$$

In the long term, the mass of pollutant tends to q/k kg.

(c) If there is, initially, no pollution in the lake, then $m(0) = 0$ and

$$m(t) = \frac{q}{k}\left(1 - e^{-kt}\right).$$

The mass of pollution increases from zero up to the steady-state level q/k kg.

(d) From Equation (S.2), with $m(t) = Vc(t)$, and putting $kV = r$ gives

$$c(t) = \left(c(0) - \frac{q}{r}\right)e^{-kt} + \frac{q}{r}.$$

This is the corresponding mathematical model for pollutant concentration.

Section 2

2.1 **(a)** When a car skids, its tyres may leave marks on the road. The purpose of the model is to find a method of using the lengths of such skid marks to work out how fast the car was travelling when it began to skid. This information may be useful to the police after a road accident, when they try to find out what happened.

(b) The test car provides data in addition to the length of the skid marks in the accident. These data allow a direct comparison to be made between the length of the skid marks in the accident, for which the speed of the car is unknown, and the length of skid marks made by a car whose speed at the onset of the skid is known. In effect, the data for the test car are used to estimate the deceleration of the car involved in the accident while it was skidding. This is possible because the conditions under which the test takes place are similar to those for the accident, apart from the speed of the car. The gradient and state of the road surface, the condition of the car's tyres, and other circumstances of the accident have a significant effect on the skid. They may vary widely from accident to accident. It is more reliable to reproduce the conditions in a test than it is to make allowance for them in a table.

(c) The relationship required is one that gives the speed of the vehicle at the onset of the skid in terms of the length of the skid marks. The longer the skid marks, the faster the car was travelling when it began to skid (all other things being equal), so the speed is an increasing function of the length of the skid marks.

(d) The key variables are the original car's initial speed, u_{car}, and the length of its skid, x_{car}. The data, which are the corresponding quantities for the test car, are represented by the symbols u_{test} (initial speed) and x_{test} (length of skid). All of the symbols appearing in the final formula have now been identified, but there are still others. Two that are used to make the calculations easier are a_{car}, the acceleration of the original car, and a_{test}, the acceleration of the test car. (The final speed is zero in each case, so there was no need to introduce symbols for the final speed of either car.) There are also the four symbols v, u, a and x that are used in the general equation (2.1) for constant acceleration.

(e) No, it does not matter in this instance. One way of seeing why is to rewrite the final formula as

$$\frac{u_{\text{car}}}{u_{\text{test}}} = \sqrt{\frac{x_{\text{car}}}{x_{\text{test}}}}.$$

The left-hand side of this equation is the *ratio* of two speeds, while the right-hand side is the square root of the *ratio* of two distances. Now the ratio of two speeds is a dimensionless number, and so has the same value whether the speeds are measured in $\text{m}\,\text{s}^{-1}$, mph or leagues per century (provided that both speeds are measured in the same units).

Likewise, the ratio of two distances is dimensionless and takes the same value whatever unit of measurement is used (again provided that both distances are measured in the same units).

The omission of units in the description of the variables is therefore not important in this case, although you are usually encouraged to state units of measurement when defining variables.

(f) The model that underlies the whole discussion is that of the motion of a particle moving in a straight line with constant acceleration (which you may have met in MST209 *Unit 6*). The formula required here is that relating the initial and final velocity, acceleration and position, namely

$$v^2 = v_0^2 + 2a_0x,$$

which becomes the same as Equation (2.1) when u is substituted for v_0 and a for a_0.

(g) The assumption of particle motion in a straight line is not mentioned. (It may be reasonable, based on the police's knowledge of skids and the record left by skid marks, but it is not mentioned explicitly as an assumption.)

The assumption of constant acceleration is not mentioned either.

Two other assumptions are mentioned explicitly, in the second paragraph. The first is that 'the frictional forces between the surfaces are not dependent upon the speed of the car, only upon the mass of the car, the condition of the road and the type of surface'. The second is that the frictional force is 'proportional to the mass of the car, as are any accelerating or decelerating forces due to gravity'. From these two assumptions it is deduced that 'the decelerations of the original car and the test car are the same' (although this is restated as an assumption shortly after Equation (2.1)).

Implicit in what follows, though not obvious from any part of the example, is the assumption that the only motive or resistive forces acting on a car are friction and 'accelerating and decelerating forces due to gravity' (that is, weight). Thus, for example, it has been assumed that air resistance can be ignored. It could be argued that the lack of mention of forces other than friction and weight amounts to an assumption that no other motive or resistive forces are acting. In sum, then, the only forces assumed to be acting on a car are friction, its weight and, of course, the normal reaction.

Another assumption, implicit in carrying out the test skid, is that the road and tyre conditions are the same for both cars. (It is not clear whether the masses of the cars are assumed to be the same.)

(h) In the light of the answer to part (g), the only forces that are assumed to be acting on a car are its weight $\mathbf{W}$, the friction $\mathbf{F}$ and the normal reaction $\mathbf{N}$. Since the motion is assumed to be in a straight line only two axes are needed. The direction of motion is horizontal, because the road is flat. Modelling the car as a particle of mass m gives the following force diagram.

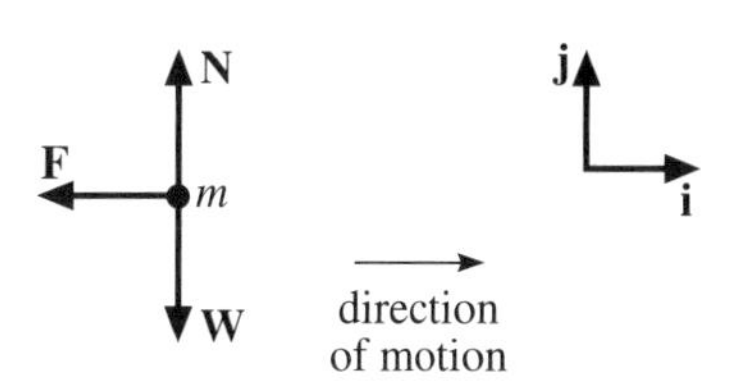

Newton's second law gives

$$\mathbf{F} + \mathbf{W} + \mathbf{N} = m\mathbf{a},$$

where $\mathbf{a} = a\mathbf{i}$ is the acceleration of the car. Resolving in the $\mathbf{i}$-direction gives

$$-|\mathbf{F}| = ma. \tag{S.3}$$

Resolving in the $\mathbf{j}$-direction gives

$$-|\mathbf{W}| + |\mathbf{N}| = 0. \tag{S.4}$$

Now $\mathbf{W} = -mg\mathbf{j}$, so $|\mathbf{W}| = mg$. Hence, by Equation (S.4), $|\mathbf{N}| = |\mathbf{W}| = mg$. Since the car is skidding (sliding),

$$|\mathbf{F}| = \mu'|\mathbf{N}|,$$

where μ' is the coefficient of sliding friction. Hence $|\mathbf{F}| = \mu' mg$. Using Equation (S.3), it follows that $ma = -\mu' mg$, leading to

$$a = -\mu' g. \tag{S.5}$$

For any given set of road and tyre conditions, μ' is constant. Thus Equation (S.5) justifies the assumption of constant acceleration (that is, deceleration).

(Note that if air resistance, or any other force that depends on velocity, were included, then the acceleration a would not be constant. Hence the assumption of constant acceleration implies that all such forces can be ignored.)

(i) Since the road and tyre conditions are assumed to be the same for both cars, then μ' is the same for both. Hence, by Equation (S.5), a is also the same for both, so the 'assumption' that the decelerations of the two cars are the same is justified.

(Note that m does not appear in Equation (S.5). Hence the conclusion of equal decelerations is independent of the masses of the two cars, which do not therefore need to be taken into account.)

(j) The model does apply to an accident on a slope, provided that the slope does not vary. To see this, consider a car skidding down a constant slope which is at an angle θ to the horizontal. (The analysis for a car skidding up a slope is similar, and results in similar conclusions, but it is not given here.) The force diagram is as follows.

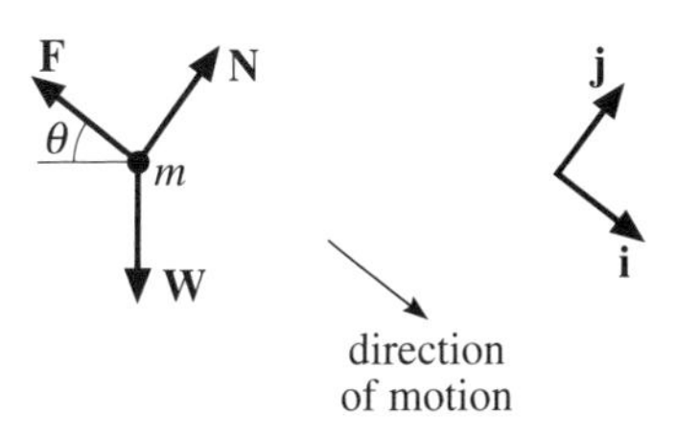

Newton's second law gives

$$\mathbf{F} + \mathbf{W} + \mathbf{N} = m\mathbf{a}.$$

Resolving in the **i**- and **j**-directions gives

$$-|\mathbf{F}| + mg\sin\theta = ma, \quad -mg\cos\theta + |\mathbf{N}| = 0.$$

Using $|\mathbf{F}| = \mu'|\mathbf{N}|$,

$$ma = mg\sin\theta - \mu' mg\cos\theta,$$

so that

$$a = (\sin\theta - \mu'\cos\theta)g.$$

Hence, provided that the slope is constant (so that θ is constant) and the road and tyre conditions are constant (so that μ' is constant) then a is constant and is the same for both cars. So the model does apply to skids along a constant slope.

(Assume here that $\tan\theta < \mu'$, so that $a < 0$. Otherwise the car will never stop!)

(k) (i) The model will apply to a Rolls-Royce, provided that the test car has similar tyres. The size and weight of the crashed car (let alone the price) are not relevant, so long as the coefficient of sliding friction can be duplicated in the test.

(ii) A crash into a headwind will be covered by the model, provided that the effects of air resistance are still ignored (this assumption was referred to in part (g)). Otherwise, the headwind will increase the air resistance force, which depends on the speed of the car relative to the air. If air resistance is included in the model, then the assumption of constant deceleration will no longer apply.

(iii) If the road is wet at the time of the crash, but dries before the test, then the model will not apply. The assumption that the acceleration in the test is the same as that in the crash will not be valid. The coefficient of sliding friction for a wet road is different from (less than) that for a dry road.

(l) The formula $u_{\text{car}} = \sqrt{-2a_{\text{car}}x_{\text{car}}}$ does predict that u_{car} is an increasing function of x_{car}, as expected.

However, by itself this is not enough to solve the problem, because there is no direct way of finding the value of a_{car}. The test overcomes this difficulty, and yields a value for $a_{\text{car}} = a_{\text{test}}$, provided that the assumptions are satisfied.

(m) If the skid marks of the original car are four times as long as those of the test car, then $x_{\text{car}}/x_{\text{test}} = 4$, so that $u_{\text{car}} = 2u_{\text{test}}$. This means that the original car was travelling at 80 mph when it began to skid, and so was exceeding the 70 mph speed limit.

(n) The instructions given to a police officer, who is going to the scene of an accident for the first time, might take the following form.

1. If there has been a skid that has left a mark on the road, measure the length of this mark.
2. Check that the tyres on the car involved in the accident are in a comparable state of wear to those on the test car, and that the slipperiness of the road has not altered significantly since the accident (for example, due to a change in the weather).
3. With a clear road, drive the test car (at a safe speed) towards the spot where the accident took place, and induce a skid. Note the initial speed of the test car and measure the length of its skid.
4. To calculate the speed of the original car when it began to skid, divide the length of the original skid by the length of the test skid. Then take the square root of the result, and multiply this by the initial speed of the test car (in mph). The answer will be an estimate of the speed, in mph, of the original car when it started to skid.

(o) The model has been based on the assumption that the final speed of the original car is $v_{\text{car}} = 0$. If this is not the case then, provided that the test still gives the correct value for the deceleration,

$$u_{\text{car}} = \sqrt{v_{\text{car}}^2 - 2a_{\text{car}}x_{\text{car}}} > \sqrt{-2a_{\text{car}}x_{\text{car}}}.$$

Hence the actual speed of the crashed car at the start of its skid would have been greater than that estimated using the model.

2.2 **(a)** Compound symbols for variables, such as u_{test} and u_{car}, immediately indicate what they represent. It is also useful to have the same base symbol for all the variables of one type, as in this case where u stands for speed. However, compound symbols are more cumbersome to write and manipulate.

Single symbols for variables, such as u and U, are easier to work with, but there may be difficulty in remembering what they represent. A compromise would be to use single-letter suffices, for example u_{t} and u_{c}; there is some danger of forgetting what these represent, but less so than with symbols such as u and U.

(The use of compound symbols without suffices, such as uc or $ucar$, is not recommended, because it is unclear whether this is intended to represent a single variable or whether it stands for a product of variables, each of which is represented by a single symbol.)

(b) Parameters do not change their value during the process. Mathematical models often involve differential equations, and knowing which symbols are constant in such equations is important when considering their solution.

(c) Although in practice there seems to be an upper limit of 1 for the coefficient of friction between tyre and road surface, there is no point in stating this as an assumption. One of the reasons for the model is to avoid determining the coefficient of friction. This is in any case a statement about a data value and not an assumption.

(d) There are four variables in the final formula: u_{car}, u_{test}, x_{car} and x_{test}. The last three are measured and their values are used to estimate u_{car}.

To validate the model, it would be necessary to measure corresponding values for all four variables, and to check that the predicted value for u_{car} is a reasonable approximation to the measured value.

There are no parameters in the final formula. However, an alternative view of the model (as in Solution 2.1(d)) is that data values for u_{test} and x_{test} are required in order to obtain a value for the parameter $a = -u_{\text{test}}^2/(2x_{\text{test}})$, which is constant for the given road conditions. This parameter value and the measured value of x_{car} are then used to find $u_{\text{car}} = \sqrt{-2ax_{\text{car}}}$.

(e) This is a subjective assessment and depends on the reader. The report could have been improved by the inclusion of a diagram, a table of variables and parameters, and a list of the assumptions on which the model is based. The final sentence is intuitively correct, but is it confirmed by the mathematics? On the whole, this report sets out what it wants to do fairly well, but requires a better format.

Section 3

3.1 (a) Features such as convenience of handling and stacking must, presumably, be taken into account. However, the most important thing, from the manufacturer's point of view, is surely to minimise the cost of making each can.

In terms of the shape of a cylindrical can, this means that the 'best' shape will be that which uses as little tin plate as possible, while meeting the requirement that the can is to hold a specified quantity of baked beans.

(b) A possible statement is as follows:

> The problem is first to find a formula for the area of tin plate needed to make a cylindrical can for containing a specified volume of food, where the formula is in terms of the height and radius of the can. This formula will then be used to find the dimensions of the can for which the area of tin plate required is a minimum.

Note that volume, rather than weight (or, more properly, mass) is used to specify the quantity of food here because, as mentioned in the main text, there is a degree of uniformity in the size of cans containing a range of different weights of foodstuff. The '400 gram' cans mentioned, on the evidence of supermarket shelves, may contain anything between 390 and 440 grams of food.

3.2 As it will be difficult to estimate the amount of tin plate used in the seams of the can, assume (at least for a first model) that the seams use a negligible amount of tin plate. It is also reasonable to assume that there is no wastage of tin plate: any 'leftover bits', resulting from cutting circular pieces for the ends of a can from a sheet of tin plate, are assumed to be recycled at negligible cost. Assume also that the cans are perfectly smooth cylinders; there are no corrugations on the surface, nor any ring pulls.

You might have thought of assuming that the cost of a tin can is proportional to the area of tin plate used, and that the manufacturing cost is negligible. These would have been appropriate if the purpose of the model had been stated in terms of minimizing a financial cost, which is the ultimate aim, but they are essentially built into the problem statement in Exercise 3.1(b).

3.3 The key variables are the area A of tin plate used in the manufacture of a can, the height h of the can, and the radius r of each end. In SI units, the linear dimensions would be in metres. However, the metre is rather too large a unit here, while the millimetre is too small, so the centimetre seems to be the best choice. Then A must be measured in cm^2, with each of h and r in cm.

The specified volume V (in cm^3) is a parameter of the model.

3.4 A possible diagram is as follows.

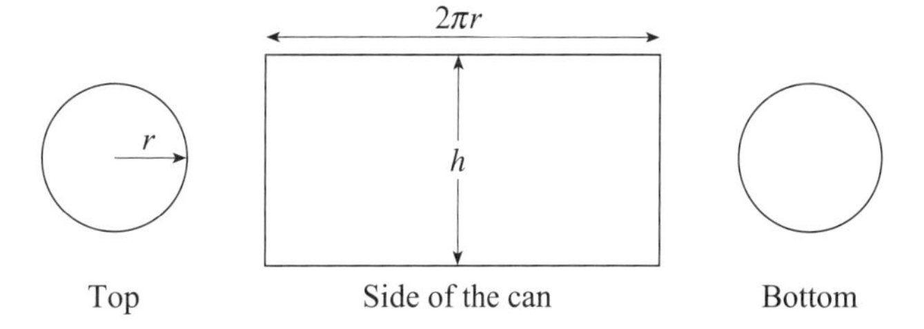

3.5 The can is made up of one rectangle (which forms the curved body of the can), of sides h and $2\pi r$, and two circular ends, each of area πr^2. Hence the area of tin plate used to make a can is

$$A = 2\pi rh + 2\pi r^2. \tag{S.6}$$

This formula relies on the assumptions that no tin plate is used for the seams, that there is no wastage of tin plate and that the surfaces of the can are perfectly smooth.

The volume of the can is given by $V = \pi r^2 h$. To express A in terms of r (and V, but not h), write $h = V/(\pi r^2)$ and then eliminate h from Equation (S.6), to obtain

$$A = 2\pi r\left(\frac{V}{\pi r^2}\right) + 2\pi r^2 = 2\left(\frac{V}{r} + \pi r^2\right). \tag{S.7}$$

(Note, in passing, that this equation is dimensionally consistent and that the area will always be positive.)

3.6 All of the assumptions have been used in the formulation. This is a useful check to perform; if any assumptions are not used in the derivation of the model, then their presence is questionable.

3.7 All of the variables and parameters that were defined have been used in the derivation. This is another useful check to perform; if any defined parameters or variables are not used in the derivation of the model, then their presence is questionable.

3.8 To solve the problem specified in Exercise 3.1(b), find the value(s) of r for which A (as given by Equation (S.7) in Solution 3.5) is a minimum. In fact, it is easy to sketch the graph of A against r by noting that it is the sum of a quadratic function and a multiple of $1/r$; the graph looks like this.

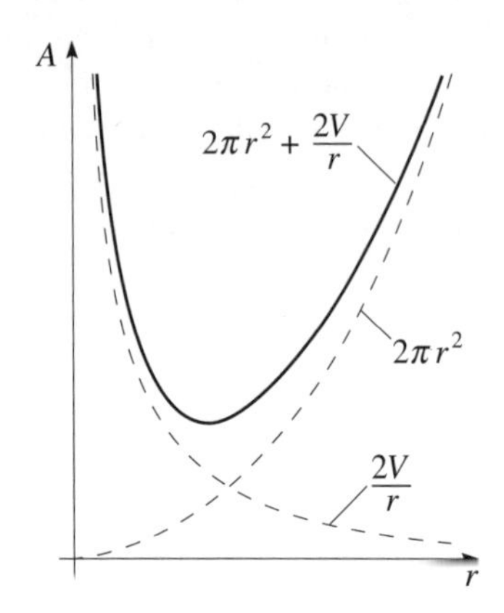

There appears to be just one local minimum, which will provide the solution to the problem.

From Equation (S.7),

$$A = 2\left(\frac{V}{r} + \pi r^2\right),$$

where V is a constant. Thus

$$\frac{dA}{dr} = 2\left(-\frac{V}{r^2} + 2\pi r\right),$$

and so $dA/dr = 0$ where

$$r^3 = \frac{V}{2\pi}.$$

(It is clear from the graph above that this gives a minimum; but to confirm this, note that

$$\frac{d^2A}{dr^2} = 4\left(\frac{V}{r^3} + \pi\right),$$

which is positive for all $r > 0$.)

Thus the minimum area, for a specified volume V, is obtained when

$$r = \left(\frac{V}{2\pi}\right)^{1/3} \quad \text{and} \quad h = \frac{V}{\pi r^2} = \left(\frac{4V}{\pi}\right)^{1/3}.$$

The corresponding expression for the minimum area is

$$A_{\text{min}} = 2V\left(\frac{2\pi}{V}\right)^{1/3} + 2\pi\left(\frac{V}{2\pi}\right)^{2/3} = 3\left(2\pi V^2\right)^{1/3}.$$

3.9 **(a)** With V fixed, the expectation is that $A \to \infty$ as $r \to 0$ (that is, the can becomes long and thin). The model (Equation (S.7) in Solution 3.5) predicts this.

(b) With V fixed, the expectation is that $A \to \infty$ as $r \to \infty$ (that is, the can becomes short and fat). This also is predicted by the model.

(c) The solution to the model predicts that the minimum area increases as the volume is increased, which agrees with common sense.

3.10 **(a)** The minimum area of a cylindrical can, of specified volume V, is given by Solution 3.8 to be $3\left(2\pi V^2\right)^{1/3}$. In this case its height is $h = (4V/\pi)^{1/3}$ and the radius of either end is $r = \left(V/(2\pi)\right)^{1/3}$, for which the shape of the can satisfies $h/r = 2$. This means that the height of the can is equal to the diameter of its ends, so that if seen from the side it will appear square.

(b) In practice, cans are rarely the shape indicated in part (a). Most of them are taller than they are wide.

3.11 **(a)** Any of the assumptions in Solution 3.2 could be behind the discrepancy, as could other factors that have not even been considered (such as convenience of handling and stacking). However, since it is relatively easy to relax the assumption of no wastage when cutting out the tin plate, this is pursued in part (b) below.

(b) If each circular end for a can is cut from a square of tin plate of side $2r$, then the formula for the area of tin plate used becomes

$$A = \frac{2V}{r} + 8r^2.$$

The minimum now occurs where

$$\frac{dA}{dr} = -\frac{2V}{r^2} + 16r = 0, \quad \text{that is,} \quad r^3 = \tfrac{1}{8}V.$$

This gives $r = \frac{1}{2}V^{1/3}$ and $h = 4V^{1/3}/\pi$. The corresponding minimum area is $A_{\text{min}} = 6V^{2/3}$, and the ratio of height to radius becomes

$$\frac{h}{r} = \frac{8}{\pi} \simeq 2.55.$$

This gives a shape more like those of actual cans, but it is still rather short and fat; the corresponding ratio for an actual can is about 3.10. Perhaps a revised model can approach the actual ratio more closely by taking the seams into account too.

This analysis suggests that there is money to be saved in changing the shape of tin cans. Alternatively, it could be argued that the cost of the can is of secondary importance. Marketing may have an important role to play in the design of the cans, since taller cans require bigger labels.

3.12 The length of metal to be soldered is given by

$$L = 4\pi r + h.$$

Using the formula $V = \pi r^2 h$,

$$L = 4\pi r + \frac{V}{\pi r^2}.$$

To minimize this length, put $dL/dr = 0$ to obtain

$$4\pi - \frac{2V}{\pi r^3} = 0, \quad \text{that is,} \quad r^3 = \frac{V}{2\pi^2}.$$

Note that $d^2L/dr^2 = 6V/(\pi r^4) > 0$, so that this will give a minimum.

This leads to $r = (V/2\pi^2)^{1/3}$ and $h = (4\pi V)^{1/3}$. The corresponding minimum length is $L_{\text{min}} = 3(4\pi V)^{1/3}$, and the ratio of height to radius is $h/r = 2\pi$.

Thus, to minimize the amount of soldering required, the ratio of height to radius should be $2\pi \simeq 6.28$. This is approximately twice the observed ratio and would result in tall thin cans.

It is likely that the cost of manufacturing the can will take into account both the cost of the tin plate and the cost of the soldering process, in which case the optimal ratio lies somewhere between 2 and 2π.

3.13 **(a)** Since 400 ml is the same as 400 cm^3, the corresponding optimal radius is

$$r = \left(\frac{V}{2\pi}\right)^{1/3} = \left(\frac{200}{\pi}\right)^{1/3} = 3.993\,\text{cm}.$$

(b) Taking $V = 404\,(\text{cm}^3)$, the optimal radius is now calculated to be 4.006 cm. The percentage change in the radius is $100(4.006 - 3.993)/3.993 = 0.33$. The predicted radius of the can is thus not sensitive to changes in the value of the volume, since the resulting percentage change in the radius is less than the percentage change in the volume.

(c) The measurement of the volume of tin cans need not be done too accurately, for the purposes of estimating the optimal dimensions of the can.

3.14 Possible key phrases are: cylindrical can of fixed volume; area; no cost of soldering; only cost is the tin plate used; minimize the area.

A possible description might read:

> This model considers a cylindrical can of fixed volume and the area of tin plate used in its construction. This area, assumed to be proportional to the total cost of manufacture, is expressed as a function of the radius. This function is differentiated to find the radius for which the area is a minimum.

Section 4

4.1 The following assumptions can be made:

(a) the change in heat energy of the tea is proportional to the change in its temperature and to the mass of the tea (as explained in MST209 *Unit 15*);

(b) all heat losses due to radiation are ignored;

(c) all heat losses due to conduction and convection through the sides and bottom of the cup are ignored;

(d) in the steady state, the rate of heat transfer by convection at a surface between a fluid and another substance is proportional to the difference between the temperature of the surface and the temperature of the substance, and proportional to the area of the surface;

(e) the tea, apart from a thin layer close to its surface with the air, has a uniform temperature;

(f) the surrounding air, apart from a thin layer close to the surface of the tea, has a uniform constant temperature;

(g) over a short time interval, the steady-state model for the rate of heat transfer by convection provides a good approximation to the rate of heat energy loss, and this approximation improves as the time interval becomes shorter;

(h) changes in the heat energy of the material of the cup are ignored;

(i) the surface of the tea within the cup is a disc.

4.2 The variables are: t, $\Theta(t)$, δE, $\Theta_{\text{sur}}(t)$ and $q(t)$. The rest are parameters.

4.3 It is better to have a common symbol for all related variables, and to distinguish them from each other by subscripts.

Using $\Theta_{\text{sur}}(t)$ instead of Θ_{sur} emphasizes that it is a function of t and not a parameter. However, the functional notation may sometimes be omitted, for convenience. (This also applies to other dependent variables.)

The variable name $\Theta_{\text{s}}(t)$ may also be used; it is shorter to write than $\Theta_{\text{sur}}(t)$, but its meaning may not be recognized so easily.

4.4 Rearranging the two equations, the temperature differences are

$$\Theta - \Theta_{\text{sur}} = \frac{q}{h_{\text{tea}}A} \quad \text{and} \quad \Theta_{\text{sur}} - \Theta_{\text{air}} = \frac{q}{h_{\text{air}}A}.$$

In order to eliminate Θ_{sur}, these two equations are added to obtain

$$\Theta - \Theta_{\text{air}} = \frac{q}{A}\left(\frac{1}{h_{\text{tea}}} + \frac{1}{h_{\text{air}}}\right).$$

Hence

$$q = UA(\Theta - \Theta_{\text{air}}), \quad \text{where} \quad U = \left(\frac{1}{h_{\text{tea}}} + \frac{1}{h_{\text{air}}}\right)^{-1}.$$

4.5 Yes, all the assumptions have been used. This is a useful check to make, to ensure that there are no redundant assumptions.

4.6 **(a)** If $\Theta > \Theta_{\text{air}}$, then the temperature of the tea would decrease, that is, $d\Theta/dt < 0$. If $\Theta < \Theta_{\text{air}}$ (e.g. iced tea), then the expectation is that $d\Theta/dt > 0$. If $\Theta = \Theta_{\text{air}}$, then the temperature of the tea should not change, so that $d\Theta/dt = 0$. Furthermore, the magnitude $|d\Theta/dt|$ of the rate of change of temperature would increase as a result of any increase in the magnitude $|\Theta - \Theta_{\text{air}}|$ of the temperature difference.

(b) Equation (4.7) says that the rate of change of temperature, $d\Theta/dt$, is proportional to the temperature difference $\Theta - \Theta_{\text{air}}$, with a *negative* constant of proportionality. So this equation is consistent with all the statements in part (a).

4.7 The differential equation can be solved by using either the integrating factor method or the separation of variables method (both described in MST209 *Unit 2*). Using the integrating factor method, first rewrite the equation in the form

$$\frac{d\Theta}{dt} + \lambda\Theta(t) = \lambda\Theta_{\text{air}}.$$

The integrating factor is $p = \exp(\int \lambda\, dt) = e^{\lambda t}$. Multiplying the differential equation by this factor and then integrating,

$$e^{\lambda t}\Theta(t) = \int \lambda\Theta_{\text{air}} e^{\lambda t}\, dt = \Theta_{\text{air}} e^{\lambda t} + C,$$

where C is an arbitrary constant. Rearranging,

$$\Theta(t) = \Theta_{\text{air}} + Ce^{-\lambda t}.$$

Putting $t = 0$ and using the initial condition $\Theta(0) = \Theta_0$ gives $C = \Theta_0 - \Theta_{\text{air}}$. Hence the required particular solution of the differential equation is

$$\Theta(t) = \Theta_{\text{air}} + (\Theta_0 - \Theta_{\text{air}})e^{-\lambda t}.$$

4.8 Since $\lambda = UA/(mc)$,

$$U = \frac{\lambda mc}{A} = \frac{3.403 \times 10^{-4} \times 0.25 \times 4190}{0.004\,07} = 87.55\,\text{W}\,\text{m}^{-2}\,\text{K}^{-1}.$$

Also

$$\frac{1}{U} = \frac{1}{h_{\text{air}}} + \frac{1}{h_{\text{tea}}},$$

so that

$$\frac{1}{h_{\text{air}}} = \frac{1}{U} - \frac{1}{h_{\text{tea}}} = \frac{1}{87.55} - \frac{1}{500} = 0.009\,422.$$

Hence $h_{\text{air}} = 106.1\,\text{W}\,\text{m}^{-2}\,\text{K}^{-1}$.

Therefore, the experimental value for λ would arise from a U-value for the convective heat transfer from the tea to the air of $87.6\,\text{W}\,\text{m}^{-2}\,\text{K}^{-1}$, which corresponds to a convective heat transfer coefficient from the surface to the air of $106\,\text{W}\,\text{m}^{-2}\,\text{K}^{-1}$.

4.9 **(a)** In either case, $h_{\text{air}} = 10$. If $h_{\text{tea}} = 50$, then $U = 8.3\,\text{W}\,\text{m}^{-2}\,\text{K}^{-1}$, whereas if $h_{\text{tea}} = 1000$ then $U = 9.9\,\text{W}\,\text{m}^{-2}\,\text{K}^{-1}$.

(b) In either case, $h_{\text{tea}} = 500$. If $h_{\text{air}} = 2$, then $U = 2.0\,\text{W}\,\text{m}^{-2}\,\text{K}^{-1}$, whereas if $h_{\text{air}} = 25$ then $U = 24\,\text{W}\,\text{m}^{-2}\,\text{K}^{-1}$.

(c) The value of U is much more sensitive to changes in the value of h_{air} than to changes in the value of h_{tea}, so this will also be the case for the estimate for T.

4.10 A better value for h_{air} might lead to a better estimate for T, the time taken for the tea to cool. In fact, as you saw in the solution to Exercise 4.8, a value of $106\,\text{W}\,\text{m}^{-2}\,\text{K}^{-1}$ for h_{air}, based on the experimental data, would provide a good estimate for T. However, this value is well outside the given range 2–25 for convective heat transfer coefficients for gases, so the model seems unlikely to be a valid one.

Furthermore, even if there is a better estimate for h_{air} in the range 2–25, the model would still predict an exponential relationship between temperature and time. So the model would still not be satisfactory, as the data show that this relationship is not quite exponential.

4.11 The two most simple shapes that resemble a cup are a cylinder and a hemisphere. For the former, an extra parameter, the height of the cup, is required.

Although either could be pursued, the model to be proposed here will assume a cylindrical shape.

4.12 The value of λ will be larger in the revised model, so that the time predicted for the tea to cool to a drinkable temperature will be less. Hence the revised model is likely to be more accurate than the first model. However, the solution is still of the same form as for the first model and so will not address the qualitative deficiencies of that model.

4.13 With $h_{\text{air}} = 32\,\text{W}\,\text{m}^{-2}\,\text{K}^{-1}$,

$$U = \left(\frac{1}{h_{\text{tea}}} + \frac{1}{h_{\text{air}}}\right)^{-1} = \left(\frac{1}{500} + \frac{1}{32}\right)^{-1} = 30.075$$

and

$$U_{\text{side}} = \left(\frac{1}{h_{\text{tea}}} + \frac{b}{\kappa} + \frac{1}{h_{\text{air}}}\right)^{-1} = \left(\frac{1}{500} + \frac{0.003}{1.5} + \frac{1}{32}\right)^{-1} = 28.369,$$

so that

$$\lambda = \frac{UA + U_{\text{side}}A_{\text{side}}}{mc} = \frac{30.075 \times 0.004\,07 + 28.369 \times 0.0139}{0.25 \times 4190} = 0.000\,494.$$

From Equation (4.9), the time taken for the tea to become drinkable is

$$T = \frac{1}{\lambda}\ln\left(\frac{\Theta_0 - \Theta_{\text{air}}}{\Theta_T - \Theta_{\text{air}}}\right) = 781.$$

Hence the given value of h_{air} leads to a predicted cooling time of about 13 minutes, which is very close to the experimental value.

This value for h_{air} is in the range for forced convection, and so the model could only be used if forced convection was taking place. It is possible that there was forced convection when the experiment was performed, so the model with this value for h_{air} may be satisfactory for finding the cooling time, but it would need a further experiment to test this. However, the model remains unsatisfactory for the purpose of discovering how, qualitatively, the temperature of the tea cools over time.

4.14 Taking radiative heat transfer into account, the differential equation would be of the form

$$\frac{d\Theta}{dt} = -\lambda(\Theta(t) - \Theta_{\text{air}}) - a(\Theta(t)^4 - \Theta_{\text{air}}^4),$$

where a is some parameter and $\Theta(t)$ and Θ_{air} are measured in kelvins. The additional term would have most effect when the difference between $\Theta(t)$ and Θ_{air} is greatest, and a lesser effect as the difference reduces. Including radiative heat transfer, therefore, could help to explain the qualitative difference between the models and the experimental data.

For the smallish temperature difference present in this model, it is good enough to replace the fourth power function by a straight-line approximation. The effect would be to increase the value of λ, and comments similar to those for Solution 4.12 would then apply.

4.15 **(a)** Since convection at the sides of the cup will be included, modify Assumptions (c), (e) and (f). Since the energy in the material of the cup is being taken into account, modify Assumption (h). The new set of assumptions, with the changes highlighted in italics and omissions marked by ..., is as follows:

(a) the change in heat energy of the tea is proportional to the change in its temperature and to the mass of the tea;

(b) all heat losses due to radiation are ignored;

(c) all heat losses due to conduction ... through the sides and bottom of the cup are ignored;

(d) in the steady state, the rate of heat transfer by convection at a surface between a fluid and another substance is proportional to the difference between the temperature of the surface and the temperature of the substance, and proportional to the area of the surface;

(e) the tea, apart from a thin layer close to its surface with the air *and to its surface with the sides of the cup*, has a uniform temperature;

(f) the surrounding air, apart from a thin layer close to the surface of the tea *and to the surface of the sides of the cup*, has a uniform constant temperature;

(g) over a short time interval, the steady-state model for the rate of heat transfer by convection provides a good approximation to the rate of heat energy loss, and this approximation improves as the time interval becomes shorter;

(h) *the change* in the heat energy of the material of the cup *is proportional to the change in its temperature and to the mass of the cup*;

(i) *the cup is cylindrical in shape*;

(j) *the thickness of the cup wall is small when compared to other dimensions, so that the internal and external surface areas of the cup can be taken to be equal.*

(b) The additional variables and parameters are given in the following table.

Symbol	Definition	Units
$\Theta_{\text{cup}}(t)$	the temperature of the cup at time t	°C
m_{cup}	the mass of the cup	kg
c_{cup}	the specific heat of the material from which the cup is made	$\text{J kg}^{-1}\,\text{K}^{-1}$
δE_{cup}	the change in the heat energy of the cup during the small time interval $[t, t+\delta t]$	J
$q_{\text{teacup}}(t)$	the rate of heat transfer from the tea to the cup at time t	W
$q_{\text{cupair}}(t)$	the rate of heat transfer from the cup to the air at time t	W
A_{side}	the surface area of the sides of the cup	m^2

In addition, assume that the values of h_{tea} and h_{air} also apply at the internal and external surfaces of the cup, respectively.

(c) First consider the tea. As in the first model, over a small time interval $[t, t+\delta t]$, the change in energy is

$$\delta E = mc(\Theta(t+\delta t) - \Theta(t)).$$

During this time interval, energy is transferred to the air from the surface of the tea and to the cup through its sides. The rates of heat energy transfer are, respectively,

$$q(t) = UA(\Theta(t) - \Theta_{\text{air}}),$$

where $U = (1/h_{\text{tea}} + 1/h_{\text{air}})^{-1}$, as before, and

$$q_{\text{teacup}}(t) = h_{\text{tea}}A_{\text{side}}(\Theta(t) - \Theta_{\text{cup}}(t)).$$

Since there is no heat energy input to the tea in the interval $[t, t+\delta t]$, the input–output principle, applied to the heat energy of the tea, gives

$$\delta E \simeq 0 - (q(t) + q_{\text{teacup}}(t))\delta t,$$

so that the change in the heat energy of the tea is given by

$$\begin{aligned} &mc(\Theta(t+\delta t) - \Theta(t)) \\ &\quad \simeq -(UA(\Theta(t) - \Theta_{\text{air}}) \\ &\qquad + h_{\text{tea}}A_{\text{side}}(\Theta(t) - \Theta_{\text{cup}}(t)))\delta t. \end{aligned}$$

Dividing both sides by $mc\,\delta t$, and then letting δt tend to zero, gives the differential equation

$$\begin{aligned} \frac{d\Theta}{dt} &= -\frac{UA + h_{\text{tea}}A_{\text{side}}}{mc}\Theta(t) \\ &\quad + \frac{h_{\text{tea}}A_{\text{side}}}{mc}\Theta_{\text{cup}}(t) + \frac{UA}{mc}\Theta_{\text{air}}. \end{aligned} \tag{S.8}$$

(This differential equation describes how the temperature of the tea varies with time. It contains the unknown function $\Theta_{\text{cup}}(t)$, the temperature of the cup.)

Now apply the input–output principle to the heat energy in the cup over the time interval $[t, t+\delta t]$. The change in the heat energy of the cup in this time interval is given by

$$\delta E_{\text{cup}} = m_{\text{cup}}c_{\text{cup}}(\Theta_{\text{cup}}(t+\delta t) - \Theta_{\text{cup}}(t)).$$

During this time interval, energy is transferred from the tea to the cup through the inner sides of the cup, and from the cup to the air through the outer sides of the cup. The rates of heat energy transfer are, respectively,

$$q_{\text{teacup}}(t) = h_{\text{tea}} A_{\text{side}}(\Theta(t) - \Theta_{\text{cup}}(t))$$

and

$$q_{\text{cupair}}(t) = h_{\text{air}} A_{\text{side}}(\Theta_{\text{cup}}(t) - \Theta_{\text{air}}).$$

The input–output principle, applied to the heat energy of the cup over the time interval $[t, t + \delta t]$, gives

$$\begin{aligned}\delta E_{\text{cup}} &= \text{input} - \text{output}\\ &\simeq q_{\text{teacup}}(t)\delta t - q_{\text{cupair}}(t)\delta t,\end{aligned}$$

so that the change in the heat energy of the cup is given by

$$\begin{aligned} m_{\text{cup}} c_{\text{cup}}(\Theta_{\text{cup}}(t + \delta t) - \Theta_{\text{cup}}(t)) \\ \simeq (h_{\text{tea}} A_{\text{side}}(\Theta(t) - \Theta_{\text{cup}}(t)) \\ - h_{\text{air}} A_{\text{side}}(\Theta_{\text{cup}}(t) - \Theta_{\text{air}}))\delta t.\end{aligned}$$

Dividing both sides by $m_{\text{cup}} c_{\text{cup}} \delta t$, and then letting δt tend to zero, gives the differential equation

$$\frac{d\Theta_{\text{cup}}}{dt} = \frac{h_{\text{tea}} A_{\text{side}}}{m_{\text{cup}} c_{\text{cup}}}\Theta(t) - \frac{(h_{\text{tea}} + h_{\text{air}}) A_{\text{side}}}{m_{\text{cup}} c_{\text{cup}}}\Theta_{\text{cup}}(t) + \frac{h_{\text{air}} A_{\text{side}}}{m_{\text{cup}} c_{\text{cup}}}\Theta_{\text{air}}. \quad \text{(S.9)}$$

(This differential equation describes how the temperature of the cup varies with time. It contains the unknown function $\Theta(t)$, the temperature of the tea.)

4.16 **(a)** The following two situations for cooling of the tea are to be compared: (i) a constant air temperature $\Theta_{\text{air}} = C$; (ii) an air temperature $\Theta_{\text{air}}(t)$, with average value C over the time interval under consideration, which starts just above C and decreases slowly to finish just below C. The qualitative behaviour of the rate of cooling is likely to be very similar in these two cases.

For a smaller difference between the tea and air temperatures, the rate of cooling (heat loss) of the tea will be less. In the early stages, therefore, the tea temperature in situation (ii) will be slightly higher at a given time than the corresponding tea temperature in situation (i). However, for a given tea temperature in the later stages, the heat loss will be slightly greater in situation (ii). Depending on the precise nature of the function $\Theta_{\text{air}}(t)$ and on the length of the time interval considered, the final tea temperature for situation (ii) may or may not be lower than that for situation (i).

(b) A linear relationship for the ambient temperature in terms of time is

$$\Theta_{\text{air}}(t) = 18 - \frac{t}{90 \times 60} = 18 - 0.000\,185t.$$

(c) Incorporating this into the first revised model gives

$$\frac{d\Theta}{dt} = -\lambda(\Theta(t) - 18 + 0.000\,185t),$$

where

$$\lambda = \frac{UA + U_{\text{side}} A_{\text{side}}}{mc} \simeq 0.000\,166.$$

The integrating factor method gives the solution as

$$\begin{aligned}\Theta(t) &= \left(\Theta_0 - 18 - \frac{1}{\lambda}0.000\,185\right)e^{-\lambda t}\\ &\quad + 18 + \frac{1}{\lambda}0.000\,185 - 0.000\,185t\\ &\simeq 60.9e^{-0.000\,166t} + 19.1 - 0.000\,185t.\end{aligned}$$

Section 5

5.1 **(a)** The model is based on the assumption that the pollutant does not biodegrade or decay. The model would not be valid if the pollutant were unstable, in that it might be broken down by other elements in the water or by sunlight, for example.

(b) A significant reduction in the input of PCBs after 1975 would be expected, although there may be significant amounts of residual pollutant in the catchment area of the lake that may take a number of years before they are washed into the lake. The model is based on the assumption that all sources of new pollution cease. This assumption may be reasonable if the input of PCBs to the lake has, as a result of the ban, reduced to insignificant levels.

(c) As one of the two lakes (Lake Michigan is the other) that is not downstream of any other lake in the system, there would be no input of pollution to Lake Superior from any of the other lakes.

5.2 **(a)** The model assumes no new pollution, and PCBs were banned in 1975. Any data from before this are therefore inappropriate for comparison purposes. Since there is a time delay in the ecosystem as pollution works its way up the food chain, it is likely that the level of pollution in herring gull eggs reflects the pollution levels in the lake at an earlier time. It may thus be wise to disregard also the data point for 1975.

(b) The slight rise in the level of PCB concentration towards the end of the period covered by Table 5.2 could be due to a number of factors. One factor may be the existence of PCBs in other parts of the ecosystem. Since PCBs are stable, they will continue to be present in the environment at a significant level for many years.

5.3 Any of the assumptions, listed on page 7, could be challenged.

- Assumption (a) is that all sources of pollution cease. In reality there may continue to be new pollution for several years.
- Assumption (b) is that the pollutant does not biodegrade or decay. In reality all substances biodegrade or decay, although some do so more quickly than others.
- Assumption (c) is that the pollutant is evenly dispersed. In reality some areas of the lake, close to the sources of pollution, will have a higher concentration of pollutant than the rest of the lake.
- Assumption (d) is that water flows into and out of the lake at a constant rate. In reality there will be seasonal effects, with more water entering the lake during rainy periods and less water entering during dry periods.
- Assumption (e) is that all other gains and losses can be ignored. In reality drinking water is extracted, water leaves the lake by evaporation, and so on.
- Assumption (f) is that the volume of water in the lake is constant. In reality there will be seasonal effects, with more water in the lake during rainy periods and less water during dry periods.
- Assumption (g) is that negligible pollution flows into downstream lakes from upstream lakes (this is not required for Lakes Superior and Michigan). This appears to be unlikely.

Since the rate at which PCB concentration diminishes in Lake Superior is much faster than that predicted by the model, the likeliest candidate for re-examination is Assumption (b).

5.4 The rate of volatilization will depend on the surface area of the lake and on the concentration of pollutant near to the surface of the lake. It will also depend on the strength of the wind and on the concentration of the pollutant already in the atmosphere. It may also depend on temperature.

5.5 **(a)** Let $c_{\mathrm{B}} = c$ be the concentration of pollutant in the bay, so that $c_{\mathrm{L}} = \frac{1}{2}c$ is the concentration of pollutant in the rest of the lake. The rate of output of pollutant from the bay (in $\mathrm{kg\,s^{-1}}$) by each of the three mechanisms is as follows.

Water flow: $r_{\mathrm{B}}c_{\mathrm{B}} = 153c$

Volatilization: $pA_{\mathrm{B}}c_{\mathrm{B}} = 4872c$

Diffusion: $E\,(c_{\mathrm{B}} - c_{\mathrm{L}}) = 5500c$

Hence volatilization and diffusion are the most important mechanisms for removing pollutant from the bay. The amount of pollutant removal by water flow is very small by comparison.

(b) The relative importance of the three mechanisms will not change greatly if the concentration of pollutant in the rest of the lake is less than half that in the bay (assuming that the concentration in the bay is still given by $c_{\mathrm{B}} = c$). Even if $c_{\mathrm{L}}/c_{\mathrm{B}}$ is very small, the loss of pollutant from the bay by diffusion will at most double. (If c_{B} and c_{L} are nearly equal, then the loss of pollutant by diffusion may be small, but since the bay is in practice significantly more polluted than the rest of the lake, this scenario is unlikely to occur.)

(c) Two mechanisms, water flow and volatilization, are involved here. The rate of output of pollutant from the rest of the lake (in $\mathrm{kg\,s^{-1}}$) by each mechanism is as follows, for a pollutant concentration c_{L}.

Water flow: $r_{\mathrm{L}}c_{\mathrm{L}} = 4967c_{\mathrm{L}}$

Volatilization: $pA_{\mathrm{L}}c_{\mathrm{L}} = 66\,120c_{\mathrm{L}}$

Hence volatilization is the most important mechanism for pollutant output from the lake. (This is in line with a conclusion drawn from the first revised model.)

Index